Recovering Lost Speci

History for a Sustainable Future

Michael Egan, series editor

Derek Wall, *The Commons in History: Culture, Conflict, and Ecology*

Frank Uekötter, *The Greenest Nation? A New History of German Environmentalism*

Brett Bennett, *Plantations and Protected Areas: A Global History of Forest Management*

Diana K. Davis, *The Arid Lands: History, Power, Knowledge*

Dolly Jørgensen, *Recovering Lost Species in the Modern Age: Histories of Longing and Belonging*

Recovering Lost Species in the Modern Age

Histories of Longing and Belonging

Dolly Jørgensen

The MIT Press
Cambridge, Massachusetts
London, England

This book was set in Stone Serif and Stone Sans by Jen Jackowitz. Printed and bound in the United States of America.

Library of Congress Cataloging-in-Publication Data.
Names: Jørgensen, Dolly, 1972- author.
Title: Longing and belonging : recovering lost species in the modern age / Dolly Jørgensen.
Description: Cambridge, MA : MIT Press, [2019] | Series: History for a sustainable future | Includes bibliographical references and index.
Identifiers: LCCN 2019005609 | ISBN 9780262537810 (pbk. : alk. paper)
Subjects: LCSH: Restoration ecology--Psychological aspects. | Wildlife recovery--Psychological aspects. | Wildlife reintroduction--Psychological aspects. | Human ecology--Psychological aspects. | Human-animal relationships.
Classification: LCC QH541.15.R45 J675 2019 | DDC 639.9--dc23
LC record available at https://lccn.loc.gov/2019005609

10 9 8 7 6 5 4 3 2 1

For Marion and Lina
May the past always be present to lead you into the future

Contents

Contents

Illustrations

Foreword

Michael Egan

I teach an undergraduate course titled The History of the Future. It challenges students to reflect on how past societies imagined their futures. As a result, the course poses a series of feedback loops between overlapping and disparate pasts, presents, and futures—some real and some imagined—and makes a deliberate point of situating its students within this continuum. Or, at least, makes them interact with the past both as historical circumstances and as historical actors. The history of the future means to examine not just what happened, but also what people thought might happen. It unlocks the hopes and fears of past peoples and societies. And in broaching roads considered and not taken, alongside the ones that were taken, the history of the future attempts an honest dialogue in our present about our futures. Dolly Jørgensen's *Recovering Lost Species in the Modern Age: Histories of Longing and Belonging* engages in a similar exercise by examining how nostalgic practices inspired future-oriented actions through ecological restoration and species reintroduction. This topic has become all the more current with flirtations toward "de-extinction," DNA-sequencing-speak for resurrection. In so doing, Jørgensen's work strikes at the central pillars of this series: to take the past—and ideas about the past—seriously as a method of preparing for our environmental futures. Across four case studies Jørgensen invites deliberation on the intricate braiding of place, nature, and human emotions. By way of framing the work, I propose to think historically about the present: rather than drawing on past examples, I play with future nostalgias. Things we might lose. This is a kind of longing in itself.

∽

Any proper bestiary accords privileged space to the places its creatures inhabit. It is not enough to marvel at their fantastical qualities; fictional and real beings must come from somewhere. Elves (fictional, so far as I know) are of Germanic descent, Jorge Luis Borges tells us. According to Caspar Henderson, the "hirsute" yeti crab (real, as of 2005) prowls the Pacific-Antarctic Ridge, near R'lyeh. Place matters. Not only is place an indicator of habitat and characteristics, it is also necessary fodder for human connection and the stage for storytelling and, sometimes, mythmaking. Places are frequently defined by their bestiary. In other cases, creatures are the most celebrated inhabitants of particular places. How that connection with place intersects with the need to protect, preserve, or reintroduce these iconic species is fraught with old and new politics—and incommensurable interests.

∽

If the axolotl didn't exist, you would have to invent it. When I first read Julian Cortázar's "Axolotl," I assumed he was writing about a fictional creature. There was no other way to make sense of the creature he was describing, the one he "thought a great deal" about and the one he eventually became. "Now I am an axolotl," he wrote: a Kafka-esque metamorphosis occurring at a slower, Latin pace. But axolotls, to me, were like elves: delightfully not of this world. Except that they are of this world. What's interesting is that axolotls are salamanders that do not undergo metamorphosis, so perhaps my Kafka reference is best ignored. They never really get out of the tadpole frame of mind. And they have a prodigious capacity for regenerating lost limbs, which suggests all kinds of metaphors about resilience. Axolotls are real. They are the very famous inhabitants of the remaining wetlands and canals around Mexico City. The Aztecs revered the "water monster." The Spaniards less so, or else they didn't give it much thought. As Mexico City became Mexico City, the valley's lakes were drained, and axolotls were pushed to the periphery. More recently, pollution has become an even greater threat than habitat loss. Its numbers have suffered a precipitous decline since the beginning of the twenty-first century. Going back to Aztec times, the axolotl's cultural ties to the region are every bit as significant as their ecological ones. Mexico City's representative emoji is an axolotl. So what happens when it isn't there anymore? If the axolotl didn't exist, you would have to invent it. In the meantime, current efforts to protect and preserve the axolotl are both heroic and moving.

∞

In the influential *Uncommon Ground* collection, Richard White examined work and place and environment, using a Pacific Northwest pro-logging bumper sticker that read "Are you an environmentalist, or do you work for a living?" as a springboard.[1] During the 1990s, the spotted owl controversy pitted work against environment. Confronted with clear-cutting practices, environmentalists struggled to protect forests and wildlife by pointing to the precarious survival of the region's endangered spotted owl. Longing and belonging. But, actually, the controversy worked the other way around. Faced with a long and precipitous decline in logging jobs, the industry warned that even more jobs would be cut because of the owl's endangered status. Lines were drawn across Washington and Oregon between small sawmills and local environmentalists. The spotted owl remained (and remains) an endangered species, its numbers in decline. I saw a slightly cruder—and less poignant—variant of White's sticker on the back of a pickup truck: "Want to save trees? Wipe your ass with an owl." Except I saw it in Hamilton, Ontario, in a strip mall parking lot in 2018. I regret not lingering to see the sticker's owner. Thirty years and several thousand miles away from the original controversy, I was curious to know whether any longing or belonging featured in the decision to buy and apply the decal. I wonder how that conversation might have gone and whether he really had it in for the spotted owl, or owls in general. (I simply assume it was a "he"—and maybe that deserves its own investigation.)

How do species become endangered? What precipitates their decline? In what ways is endangerment a quantitative assessment of population over time—or a qualitative absence from places they once visited? Who counts? And why? Who misses them when they are gone? Who wants them back? Who is glad they are gone?

There are, of course, plenty of animals whose power to drive the human imagination transcends space. They become global superstars. Many of these are recorded as "charismatic megafauna." Though exceptions exist, they tend to be big and mammal. They also tend to be endangered. Playful pandas. Emaciated polar bears, clinging to emaciated polar icebergs. Whales, many of whom are fading into the realms of mythical or fictional beasts.

Several times in recent years, Japan has threatened to withdraw from the International Whaling Commission and resume commercial whaling. Japan had previously defended its right to hunt whales as part of its cultural tradition. But that tradition is fraught; in some corners that whaling tradition was constructed after World War II. This is not the only invented tradition surrounding species at risk or species that have disappeared.

∞

Safari and ecotourism. See them before they're gone.

∞

Visiting Lonesome George's mausoleum at the Charles Darwin Research Station in the Galápagos is a somber experience. Before entering, your group is ushered into an antechamber for several minutes, to control light and air entering the main room. This has the added, sobering effect of muting chatter. When the doors are opened, you enter into a relatively small, dark space. Lonesome George is behind glass, his neck craned. The dim light makes him more a shadow than a former creature. He is massive, proud, beautiful. Lonesome George was the last Pinta tortoise. He was found almost by accident in 1971. Prodigious efforts to mate him with similar species from other islands in the archipelago—after a global search of zoos that might have a Pinta tortoise—all failed. Lonesome George died on June 24, 2012, almost six years to the day after Bruno the troublesome Italian bear was killed in Europe. (Bruno's story unfolds in Jørgensen's concluding chapter.) There was some debate as to where George would be remembered. An effort to move his taxidermied remains to Quito, where more people could visit—bear witness to—the last of the Pinta tortoises, was met with vociferous opposition from Galápageños. Lonesome George belonged to the Galápagos and he should stay at home. His mausoleum is on Santa Cruz, roughly one hundred miles south of the much less trammeled Pinta Island where he came from, attached to a tortoise breeding center that was unable to salvage his species.

∞

Imagine being the last of your kind. There is something profoundly moving in the photographs that record the last member of a dying species. Sudan, the last male northern white rhinoceros, died in March 2018. The pictures on social media and in newspapers constituted a beautiful lament. Photographs themselves, of course, create a kind of nostalgia. Of course, too, as a species: we have become rather good at lamenting.

∽

The Jevons paradox has little interest in lost animals. It asserts that whereas new technologies or policies designed to increase the efficiency of resource exploitation should slow consumption, the opposite occurs: demand and consumption both rise. I suspect that a similar paradox applies to nostalgia surrounding lost animals or animals nearing extinction. We do not miss them until they are endangered or, worse, gone.

∽

"Lost" is a curious word. According to an online etymology dictionary, "lost" implied "wasted, ruined, spent in vain" as early as 1500; by the 1520s, it was understood to also mean "no longer found, gone astray." "Lost" comes from Old English *losian*, which is to "be lost, perish." As one might imagine, it is closely related to the suffix "-less," among whose early (and contemporary) meanings include "lacking," by way of the Middle English *leese*: "deprived of." And another root, *forleosan—forliasa* in Old Frisian, *farliosan* in Old Saxon, *verliesen* in Middle Dutch, and *firliosan* in Old High German—shows common ancestry between "lost" and "forlorn."

∽

"Lost" also implies a curious human-animal relationship. It is perfectly apt to accord much of the responsibility not just in losing—but also in disappearing—animals to humans and human practices. But one must beware that the human action of losing does not render total passivity to the lost animal. Even as ghosts, animals are agents in this study.

∞

According to Jan Assmann, "The past is not simply 'received' by the present. The present is 'haunted' by the past and the past is modeled, invented, reinvented, and reconstructed by the present."[2] True. One could also happily invert this statement: the past is haunted by the present.

∞

From ghosts to zombies. Homero Aridji's *Zombie Town* is a meditation on death. Drawing from nature and folklore, Aridji tells how in Central Mexico people become monarch butterflies after they have died. It is a charming image, even more so when you reflect on the peaceful majesty of these beautiful creatures as they settle at the southern terminal of their monumental migration around about the annual celebration of El Día de los Muertos. But in Aridji's telling, as fewer monarchs return, more and more of the city's dead are becoming zombies. That is the premise for his novel.

∞

The movement to reintroduce lost species makes me think that maybe we can be nostalgic axolotls, able to regrow not limbs but lost patches of our collective memories. After we invent them, that is. Now I, too, am an axolotl.

∞

I almost managed to write this whole foreword without invoking the Sixth Great Extinction as a means of universalizing the importance of Jørgensen's work and re-stressing the stakes of our contemporary environmental crisis.

∞

Bruno was a bear
Lonesome George was a tortoise
Their stories belong

Preface: Returning Back to the Beginning

The book you are about to read is not the book I intended to write—at least, not when I started doing research on restoration in 2012. I had thought that I was going to write a book about the constructions of an animal's nativeness and how that affected acceptance of reintroduction. That's the project that was funded by the Swedish Research Council Formas in 2013: "The Return of Native Nordic Fauna." But sometimes research doesn't turn out as planned. It turns out better.

I had decided literally from day 1 of the project to blog about my research on a regular basis, and I did. I wrote all the time—two posts per week the first year of the project, then one post for the next two years (you can still read that material online at http://dolly.jorgensenweb.net/nordicnature/). I had given myself this space to explore the material I was collecting in real time, as well as the leeway to make diversions into other interesting, related areas. It was in one of those sidetracks early in 2013 that I wrote about discovering de-extinction; later, I took another side alley into rewilding. I began to connect the pieces and saw that the history I was writing wasn't just about reintroduction, but also about rewilding and resurrection of species as extensions of that earlier idea.

That still didn't get me to where I am with this book; I continued to believe that the core question had to do with a species belonging or not because it was "native." But as I worked with the material, reviewing the thousands of pages of documents and photographs I amassed over the course of my research, I noticed that *belonging* wasn't the only part of the story. Instead, it was *longing* that popped off the page. I realized that all these people involved in very intensive and invasive actions to return a species to places where it was now extinct had an emotional investment. That became the narrative I wanted to explore: a story of emotions as motivations for recovering lost animals. That is finally this book.

Of course, this book wouldn't have come about without first and foremost the support of my husband and fellow scholar Finn Arne Jørgensen. He knows far more about beavers and muskoxen and passenger pigeons than anyone not writing this book should ever have to know. He's always waited patiently in museums as I took far too many pictures of stuffed animals and labels. Although the words in this book are mine, they are shaped by him helping me think through it all. I worked and talked so much about the animals in this book that my daughters, Marion and Lina, are convinced that they are my favorite animals. I don't really have the heart to tell them I prefer guinea pigs.

I owe the impetus to investigate restoration as a phenomenon to Christer Nilsson, who took a chance on hiring a historian as his RESTORE project coordinator. He gave me the space to figure out what I wanted to work on within the general framework of restoration, and I ended up developing this project. I am forever in his debt for that. The Landscape Ecology group at Umeå University was always a pleasure to work with, even as the odd humanities scholar in their midst. After my time at Umeå, I was thankful to be integrated with a great group of historians at Luleå University of Technology for two years. My move to University of Stavanger in 2017 gave me the opportunity to complete this manuscript. All these positive working environments contributed to this finished product.

I've had the opportunity to share this research in many venues over the last five years. I've presented parts of this project at conferences of the Agricultural History Society, American Society for Environmental History, and European Society for Environmental History; the Swedish Science and Technology Days; the Animal History Group London summer conference; the Animal Housing: Practical Infrastructures and Infrastructural Practices workshop; the Foreign Bodies, Intimate Ecologies workshop; the Im/mortality and In/finitude in the Anthropocene symposium; the Future of Wild Europe Early Career Researcher conference; the BIOMOT Beyond Economic Valuation conference; and the Rewilding in a Changing Europe conference. I also shared portions of this research as an invited speaker at Aberdeen University, Bath Spa University, University of Bristol, Brown University, Elvarheim Museum, Næs jernverket, Lund University, the Rachel Carson Center, and Tallinn University. I thank all the organizers of these events for giving me the chance to think through my work in progress and hear feedback about it. I also had the amazing opportunity to sit in Marianna Dudley's office at the University of Bristol for over two weeks to work on manuscript revisions, so I give special thanks to Marianna for arranging my visiting professor status.

Along the way, my research process was helped by archivists and librarians at the Jamtli archives, National Archives of Norway, Naturvårdsverket archive, Nordiska Museet archive, and Norwegian Polar Institute archive. I also owe Peder Roberts a special word of thanks for getting documents for me from the National Archives in Tromsø. Without the raw materials, writing history is impossible.

In addition, you can't do the research necessary for a project like this without money, so I thank those who were my financial backers at different times in the project: Swedish Research Council Formas, the Birgit och Gad Rausings Stiftelse för Humanistisk Forskning, Stiftelsen J. C. Kempes Minnes Akademiska fund at Umeå University, and University of Stavanger.

Finally, I want to say thanks to Michael Egan, who encouraged me to contribute to this book series, and Beth Clevenger, for believing in the project even when I was years late on delivering a manuscript. Beth may not remember it, but when we were discussing the first draft of my introduction, she encouraged me to write more like my blog, which was crisp and engaging. I took that to heart and changed the way I was telling these stories. I went back to the beginning, returning to the style I had used when first engaging with my material. My hope is that these environmental histories will demonstrate the value of reconnecting the emotions of the past to the present and the present to the future.

1 Recovering: Losing and Finding Nature in History

In June 2014, London-based street artist ATM released a poster that circulated online.[1] It was designed like a standard missing person or pet poster. "LOST" appeared in capital letters across the top. Underneath, rather than a dog or cat, boy or girl, there was an image of a beaver labeled "British Beaver Last seen 1587." Below that was written "Very attractive and much-missed member of our fauna. Reward offered for safe return: Fabulously rich wetlands," to encourage the reader to return the missing animal if it was found. The artistic piece was created as social commentary on the contemporary debates about whether beavers should be reintroduced into Britain, where they had been extinct in the wild. Beavers likely were hunted to extinction in Britain and Wales in the Middle Ages, although they may have survived unnoticed in small pockets until as late as the eighteenth century.[2] A reintroduction trial project in Scotland began in 2009 and finished in 2015. It subsequently was approved as a permanent beaver reintroduction, meaning that the beavers would not be removed. Meanwhile, a group of "renegade" beavers had been spotted living comfortably along the River Otter in Devon, even though a permit for release of beavers there had never been granted. ATM's intervention into this politically charged nature conservation foray not only highlighted public sentiment in support of beaver reintroduction, but also the nostalgia driving those feelings. The overwhelming sentiment is that something was lost, we miss it, we are looking for it, and we want it back.

ATM is not alone in these feelings about lost nature and desires for its recovery, nor is the sentiment something new. Although the main early environmental movements usually are classified as either conservationist or preservationist, a third movement coexisted with those: restoration. Looking at the big historical picture, the overarching aim of restoration is to bring back something that has been lost in a particular place's environment. The thing to be recovered may be a prior ecological structure (like undamming a river so that it returns to its prior watercourse), a function (like having soil that can grow crops), or a component (like an extinct or rare animal species). In all cases, restoration is about finding and recovering the lost: lost and found.

It's the feeling of environmental lost-ness and the potential found-ness that motivates decisions about recovering locally extinct animals like beavers. The beaver of ATM's Britain is missing, which implies that it would be there if all was right in the world. Something thus is wrong and needs to be fixed. Saying the beaver belongs in Britain is powerful because it asks for a recognition of the country as a home range, a place that beavers should be. Because they aren't there now, the implication is that humans should do something about it. We are asked to get involved, to find the lost in practical ways.

There have been many motivations for environmental restoration practices over their long and complex histories. Carolyn Merchant has situated restoration, particularly the Western desire to recreate Eden on Earth, as a primary driver in the history of human-nature relations.[3] To George Perkins Marsh, writing in the mid-1800s, restoration was "reclaiming and reoccupying lands laid waste by human improvidence or malice"; the pioneer settler was "to become a co-worker with the nature in the reconstruction of the damaged fabric which the negligence or the wantonness of former lodgers has rendered untenable."[4] As Marcus Hall has traced in his book *Earth Repair*, actions to right the wrongs of past overgrazing and deforestation were some of the first systematic attempts to restore ecological function for ecosystems to be usable again by humans.[5] These attempts to try to return an ecosystem to a prior state were completely anthropocentric; the humans wanted to be able to use the land for production.

Aldo Leopold and others working to recover the lost prairie ecosystem in Wisconsin are often cited as founding the concept of restoring environments for their own sake (ecocentric restoration),[6] but even then, humans were making the choices about what the ecosystem *should* look like, revealing the inherent anthropocentrism of restoration. In the 1980s, ecological restoration became a recognized scientific practice, institutionalized with the foundation of the Society for Ecological Restoration (SER) in 1987. SER adopted its first primer on ecological restoration in 2002, which set the tone for restoration as a scientific endeavor, rather than an ad hoc collection of practices.[7] In that document, *ecological restoration* is defined as "the process of assisting the recovery of an ecosystem that has been degraded, damaged, or destroyed."[8] Even these supposedly objective scientific attempts to return ecosystems to their past states necessarily include decisions about what counts as "degraded" and which ecosystems are preferable.

Ecological restoration is by necessity a practice that looks to the past to make these decisions.[9] Scientists and practitioners decide that an ecosystem has been degraded, damaged, or destroyed based on what the ecosystem had been at some time in the past. There is much debate within the restoration scientific community about how much historical fidelity is necessary for an action to be considered a restoration rather than another type of ecological intervention or management. Philosophers have disputed the authenticity of restored ecosystems, and ecologists have argued about whether past ecosystem configurations are viable under present and future climatic and usage scenarios.[10] Many of the working definitions of restoration leave open which past state should be used as the baseline or target, so values come into play.[11] Although North American restoration practices often have worked under the assumption that the state of nature before European settlement is desirable, European restorationists have tended to favor states incorporating extensive cultural use of the land. Although these arguments have the potential to affect the concrete restoration activities in the field, regardless of the nuances, ecological restoration is always about identifying something that used to be there but no longer is, and then acting to bring that lost thing back in some configuration. This is where the *recovery* part of the SER definition comes in. *Recovery*, defined as "the regaining or restoration to one's control or possession of a thing lost, stolen, or otherwise taken away," is the goal.[12]

In this book, I am going to use environmental histories of reintroduction, rewilding, and resurrection to situate the modern conservation paradigm of the modern recovery of nature. I will argue that the recovery of nature—identifying that something is lost and then going out to find it and bring it back—is a nostalgic practice that looks to a historical past and relies on *belonging* to justify future-oriented action. As a nostalgic practice, recovery depends on emotional responses to the lost, particularly a *longing* for recovery that manifests itself in emotions such as guilt, hope, and grief. Since the Enlightenment, there has been an idea that humans should be objective and scientific in their relationship with nature,[13] but in this study I want to show how fundamental emotions are to how modern humans relate to nonhumans, despite our reliance on scientific principles. It is an environmental history that examines the emotional motivations behind conservation actions. The implication for environmental historians and others more generally interested in the history of conservation is that emotional frameworks matter deeply in both how people mentally understand nature and how they interact physically with it.

As the chapters progress from reintroduction to rewilding to resurrection, those involved in the recovery of nature go to greater and greater lengths to bring back the past. I believe the impulse to restore nature to a previous condition, no matter how outrageous the intervention may sound at first, is tied to the modern recognition of mounting global environmental change since industrialization. In restoration practices, there is skepticism about modernization and its consequences. As Ursula Heise astutely observed in her work about species endangerment, the loss of a particular place or species "comes to stand in for the broader perception that human relationships to the natural world have changed for the worse."[14] Today we are labeling those large-scale changes as *the Anthropocene*, a move that recognizes that human effects on the planet are massive—and even geologic.[15] George P. Marsh published his book *Man and Nature* with an explanatory subtitle, Physical Geography as Modified by Human Action, back in 1864. Marsh wrote his work both to "point out the dangers of imprudence" of large-scale anthropogenic change to biological systems and "to suggest the possibility and importance of the restoration of disturbed harmonies."[16] Under these conditions, conserving what we have left often has been seen as not enough because we have so little remaining. This spurs a drive to turn the world back to a time of more abundant nature, a time before it was lost. That means that the future is irrevocably tied to conceptions of the past.

The identification of what belongs (the lost nature) and our longing (the emotional attachment to it) in the present will affect how environmental restoration practices are carried out in the future. In *Wild Ones*, Jon Mooallem observed poignantly: "In the twenty-first century, how species survive, or go to die, may have more to do with Barnum than Darwin. Emotion matters. Imagination matters. The way we see a species can impact its standing on the planet more than anything covered in ecology textbooks."[17] As we recognize that we have extensively modified environments, often for the worse, recovery will become more and more central in environmentalism. A sustainable future will depend on questioning how and why belonging and longing factor into the choices we make about what to recover.

Longing and Belonging

Disenchantment with modernity and nostalgia for a time when things were purer, cleaner, healthier, and greener is nothing new. At least since the Enlightenment, some writers have imagined an idyllic, premodern society as a counterpoint to contemporary urbanization and industrialization.[18] Pastoral painters of the nineteenth century invoked lost agricultural landscapes to invite viewers to dwell upon prior relationships with the land.[19] The eminent historian Lewis Mumford contrasted the happy peasant running through the field to his beloved in the preindustrial days to the dirty, dangerous work in the modern coal mine.[20] With nostalgia, the grass is always greener on the other side. As David Lowenthal noted, "A perpetual staple of nostalgic yearning is the search for a simple and stable past as a refuge from the turbulent chaotic present."[21]

This kind of yearning is actualized in movements to protect, and often freeze at one moment in time, cultural heritage, natural heritage, buildings, and environments. It also prompts a looking back to the past—a longing for what was before—that can bring about change. Jennifer Ladino has labeled nostalgia that brings about change, particularly as a challenge to a progressivist ethos, as *counternostalgia*—in contrast to nostalgia, which supports the status quo.[22] I agree that we need to reclaim nostalgia as a potentially productive force, but I don't think another label is necessary. Although the activists in her book go against the prevailing social sentiments and desire change, they too are nostalgic for the past, albeit for a different past than others. The people in this book also look to a different ecological past than the average person and act to bring about change for the future—an attitude I would still classify as nostalgic. That nostalgia has become a standard in modern society should come as no surprise. Malcolm Chase and Christopher Shaw identified three conditions as prerequisites for nostalgia: a secular and linear sense of time, an apprehension of the failing of the present, and available material remnants of the past.[23] Western society since at least the eighteenth century certainly fits the profile.

Although the objects of many heritage practices are cultural, nature is no less subjected to remembrance and nostalgia. Both nature and human antiquities have been protected as inheritances for future generations since the nineteenth century.[24]

From John Muir wanting to preserve the beauty of Yosemite to Henry David Thoreau musing about Walden Pond, particular landscapes stir up nostalgic emotions and become ingrained as places worth protecting. The idea that natural landscapes like mountains, canyons, and waterfalls are part of national heritage was foundational to the rise of national parks. Now the idea stretches into the UNESCO World Heritage List to claim that some nature, like the Great Barrier Reef, the Everglades, and the Okapi Wildlife Reserve of Congo, is patrimony for *all* humans.[25] In these and other cases, animal species become symbols of heritage. Animals like the North American bison, passenger pigeon, red squirrel, and polar bear have been woven into a complex relationship between cultural and natural heritage, in which the two are inexorably joined.[26]

Heritage claims that something, whether a landscape, a building, or a species, *belongs* in a place. *Belonging* in this sense means to be related or connected, to fit in a certain environment.[27] As I discuss in this book, belonging is about relationships, not just identities. In other words, belonging is constructed, negotiated, and contested through biocultural relationships rather than being a fixed category.[28] Much of the scholarship on the contested categories of native/non-native and indigenous/alien for animals and plants exposes the power of belonging as an idea.[29] "Nativeness" is negotiated through cultural preferences, defining what belongs and what does not. As Matthew Chew and Andrew Hamilton have argued, "The nativeness standard relies on two tacit conceptual transformations. The first takes nativeness to mean a taxon *belongs* where it occurs, geographically, temporally and ecologically. The second takes *belonging* to signify a morally superior claim to existence, making human dispersal tantamount to trespassing."[30] In this construction, humans can be judged to have wronged a species if they have caused it to no longer occur in a place it is believed to belong. Sometimes belongingness is based on the historical presence of a species or on genetic evidence, but not always.[31] The whole concept of species itself is a historical construction, with unclear delineations between what constitutes a species versus a subspecies and how species might differ from race.[32] The belongingness of an animal can be in opposition to prevailing biological theory, as in the case of the cane toad in Australia, which has been reviled as a destructive invasive species and subjected to violent control measures yet is kept as pet and has even been proposed as a State of Queensland icon.[33] Whether ducks or crocodiles, spiders or trees belong in a place is a contested matter.[34]

Labeling animal species as belonging or giving them names like native, exotic, or invasive creates, in Thom van Dooren's words, "valued natures."[35] These are the natures humans produce through our practices of species control, protection, integration, and rejection. The urbanization of the gray squirrel and its subsequent integration into American urban life and culture, as traced by Etienne Benson, is a superb example of how an animal can be valued as a species that belongs in a place, even if it's new there.[36] Belonging then is a cultural category. It is something determined by humans as a judgment on the appropriateness (or inappropriateness) of a species in a place. Natural scientific data may play into the decision-making process, but it is never the whole story behind the feeling that something belongs.

Our era of deep and substantial effects on the planet Earth and its biota brought about by human activities has been characterized as a time of loss: loss of biodiversity, habitats, and fauna species dominate the discourses about our changing globe. The recognition of species extinction as even a possibility is relatively new in human thought. It was only during the eighteenth century that scientists began to recognize that species could become extinct (i.e., not exist anywhere anymore)—and moreover, that such extinction could be caused by humans.[37] Since then, the rate of human-induced extinction has continued to rise, with the last two centuries witnessing the loss of numerous bird species, such as the great auk, huia, ivory-billed woodpecker, heath hen, and Carolina parakeet; mammals such as the Tasmanian thylacine, lesser bilby, Guam flying fox, and Caribbean monk seal; reptiles such as the Pinta giant tortoise; and insects like the Xerces blue butterfly and American chestnut moth.[38] Many of these species had a restricted range to begin with—often found on islands or tied to specific habitats—so they did not fare well with human modifications to the landscape, but others were victims of intentional extinction, like the thylacine and Japanese wolf.[39] Alongside dramatic complete extinctions, in many cases extinction happens at a local or regional level, with some individuals remaining elsewhere. In these situations, the loss is not complete, but the fact that the species is missing from a place is still framed as a loss for the species and/or landscape.

The loss of animal species and their environments in the modern age has brought about multiple reactions, including expressions of grief, utopian "romantic hopes for a returned and re-enchanted intimacy with nature," and nostalgia for that which no longer exists.[40] *Solastalgia*—sickness caused by loss connected to physical desolation of one's environment—has even been proposed as a modern condition.[41] Geographer Lesley Head has smartly argued that "loss and mourning have been an explicit part of biodiversity conservation debate *because* of their temporal reference point against a past baseline."[42] The past and what was lost is measured against the future and what might be gained. The "extinction studies" field has recently been founded as an interdisciplinary inquiry into how extinction interrupts the processes of time, death, and generations.[43]

From a philosophical point of view, van Dooren has argued that grief and mourning for a species on its way to extinction or already there is a vital action in a multispecies world.[44] Loss demands grief in his environmental philosophical reading. Radical environmentalists, for example, are motivated by grief over environmental loss and participate in rites of mourning.[45] Grief and mourning have been proposed by Ashlee Cunsolo and others as legitimate and potentially productive emotional responses to ecosystem loss from climate change.[46] Such reactions are not confined to humanistic thinking: one of the leading restoration ecologists, Richard Hobbs, has posited that scientists working in the fields of conservation and restoration ecology "live mostly in a world characterized by loss, and hence are either wittingly or unwittingly in a state of grief."[47] Restoration practitioners are not the only environmental scientists dealing with emotions, as evidenced by the growing number of books by naturalists and natural scientists that deal poignantly with the grief attached to the loss of species and ecosystem.[48] We need to recognize that feelings of loss can motivate past and future environmental action. Although these commentators have focused on grief, I believe feelings of loss and the desire to find can manifest themselves through other emotions as well.

Guilt, hope, and grief—the emotional frameworks covered in this book and defined in the upcoming chapters—are all possible responses to loss. These feelings determine what is identified as worth bringing back and the lengths to which people are willing to go to make the return a reality. The feeling of loss creates the framework within which the environmental actions discussed in this book (and many more currently ongoing in the twenty-first century) take place. Bringing in the emotions involved in these recovery stories is a critical analytical move because, as geographer Laura Smith points out in her study of restoration, "'emotionality' can be a powerful catalyst for bringing about environmental restoration and wider narratives of landscape change."[49]

The historical discipline has witnessed a recent turn to emotions as phenomena both to be explained and that offer explanations. The scientific study of emotions with its roots in psychology and physiology has been pushed into the realm of historical studies, notably in the work of Barbara Rosenwein, advocating the study of "emotional communities"; Peter and Carol Stearns' idea of emotional standards; and in scholarship tracing the historical development of particular emotions or the ways in which they are expressed.[50] In general, scholars working in the history of emotions acknowledge that there are some common physical responses to particular events (e.g., increased heart rate if scared or sweating if nervous)—which are often called *affect*—while simultaneously arguing that emotions, the outward expressions of feelings that may be brought on through affect, are socially constructed and socially interactive.

In this book, I am not interested in tracing the emotions tied to these animal lost-and-found stories for themselves, but rather as windows into understanding the motivations of the historical actors. Focusing on emotions allows me to move beyond discourse analysis, which, as Lyndal Roper has rightly pointed out, "offers no account of the relationship between language and psychology, and that makes it hard to explain why particular discourses might be appealing, or what the relationship is between thought and action."[51] The idea that emotions are practices— "something people experience and something they do"—allows us to see emotions in both what people say and how they act.[52] As historian of memory Alan Confino remarked, emotions are "a force giving shape to politics, society and culture, to beliefs and values, and to everyday life, institutional settings, and the processes of decision making."[53] Emotions, then, are a way to understand conservation motivations and decision processes.

On a methodological level, getting at emotions in history is not easy. In early modern history scholarship, in which a good deal of thinking about how to make the history of emotions into a practice has occurred, writers have encouraged reading a wide variety of sources for emotions: texts ranging from letters to economic ledgers to dry monastic chronicles and objects both personal and public.[54] Emotions are not found only in overtly emotional texts, but also in other types that nevertheless describe the decision-making processes or their reception, which give us insights into thought and emotion. In a historical study, the informants cannot be asked directly how they "felt" about something they did; instead, those emotions must be inferred from the traces left behind. A history of emotions also recognizes that humans do not feel just one emotion at a time, so there can be multiple layers of emotions in the same trace.

Although the historical profession has readily adopted history of emotions approaches in many areas, they have been virtually absent from environmental history. I know of two environmental historians in the midst of projects dealing explicitly with emotions: Andrea Gaynor is working on emotional attachment to Australian urban frogs and conservation, and Michael Egan is writing on toxic fear in North America.[55] Environmental and technological historian Daniel Macfarlane also has combined aesthetic preference with emotions in a short analysis of Niagara Falls.[56] Cultural geographer Yi-Fu Tuan explored the long history of fear in the landscape, but as a geographer he is particularly interested in universalizing human fear, which runs counter to a historian's approach.[57] Although there has been excellent work done on the moral ecology of conservation, which Karl Jacoby defines as a moral universe grounded in certain worldviews that shapes interaction with the environment, environmental historians have not explored where emotions fit into these moralities.[58] There is much room remaining for an environmental history of emotions.

The lack of an explicit history of emotions in environmental history is somewhat surprising given its major place in other disciplines within the environmental humanities. Both ecocriticism and environmental film studies have been inundated with studies of affect.[59] These disciplines have been particularly interested in the way that literary and artistic works can inspire environmental action and activism by exploring the intersection of biology, materiality, and cultural specificity to embed people in environments. The enjoyment, pleasure, or disgust at reading a text or watching a film is fundamentally affectual. I do not use the term *affect* in this book, but the insights from this scholarship about environmental narrative construction and emotional appeal is foundational to how I have read emotions in my historical sources and how I have reconstructed narratives deployed by my historical subjects. Following the lead of Head's book *Hope and Grief in the Anthropocene*, I look for emotions as expressed through practice. Emotions are located by analyzing actions and what historical actors said or wrote about their motivations.

Recovery Practices

Within the scope of this book, I focus on three recovery practices: reintroduction, rewilding, and resurrection. All these words start with the prefix *re-*, which basically means to do something again or revert to a previous state. *Re-* implies that we are repeating something that has been done before. To *reintroduce* is to bring an animal species back to where it used to live; to *rewild* is to revert to a state of wildness; and to *resurrect* is to rise or grow again. These *re-* words are then fundamentally historical acts pointed toward the future. They cannot help but be nostalgic in their formulation of looking to the past in service of the future.

I focus on these three recovery practices as responses that can happen after a species is gone from a particular place, rather than studying general conservation activities aimed at preserving endangered or threatened species populations that are still present.[60] I am not as concerned with the outcome for the animal species in these tales as I am with the motivations for the human actions. I am interested in probing the question, Why is it that people decide to go to such lengths to recover a locally or even globally extinct species? Before moving on to the cases, it is worth giving an overview of what these practices are and how they differ.

The concept of *reintroduction* applies when a species is locally extinct (which is technically called *extirpation*) but was known (or believed) to have lived in a specific location in the past, so individuals are brought to the locale from a surviving population elsewhere. There is a debate about how far back in time a species can last have existed in a place and still label the effort as reintroduction. Some scientists advocate looking back in time only within two to three hundred years (the modern industrial age); others include any time in the historical written record; still others include any time in the archeological record.[61] In general, though, contemporary reintroduction projects that are not also labeled as *rewilding* focus on recently extinct animals.

In chapter 2, I tell the history of the first systematic beaver reintroduction project in the world. It took place in Sweden in the 1920s and 1930s, becoming a model for reintroductions throughout Europe. There were earlier conservation breeding projects, but none was as radical as the beaver project. The American bison breeding project, which was spurred on by naturalists such as William Hornaday and Teddy Roosevelt, released bison into fenced reserve ranges beginning in 1907, which was earlier than the Swedish beaver reintroduction, but it took place on protected public land.[62] In Europe, a small number of Alpine ibex also were reintroduced into the Swiss National Park in 1920.[63] In the Swedish case, beavers were set out in the wild outside of protected habitats; their new homes were regular lakes and streams in the countryside. The beaver reintroduction project in Sweden is the greatest success story in the history of animal reintroductions worldwide: beaver went from extinct to an estimated population of one hundred thousand in the country, nearly completely regaining its former range in less than eighty years.

Looking into the writings of those who spearheaded the beaver reintroduction, I found that feelings of guilt were a main motivation for the project. The beaver had been lost from Swedish nature directly because of humans systematically hunting it for medicine, fur, and meat. Stories about the beaver's local extinction fixated on human culpability, creating a sense of longing for recovery to make amends for historical human wrongs. I explore how this longing led to tangible actions to return the beaver to Sweden. There is an element of redemption, making right what had been made wrong, in the beaver's recovery—but that redemption is of the humans involved more than the beaver.[64]

The fastest growing conservation idea so far in the twenty-first century has been rewilding. At its core, rewilding is about bringing back species and ecosystems eradicated by humans over our long history, and it has appeared in both scientific and activist writing. Unlike reintroduction projects that focus on a single species that is brought back specifically to conserve the species or to restore a particular function that the species performs, rewilding focuses on the larger landscape level, often with the idea that the landscape should be returned to some kind of state prior to human intervention. Often, rewilding initiatives propose that the landscape and its fauna should be returned to a state at the end of the Pleistocene (approximately 11,700 years ago).[65] Despite this basis in the past, rewilding advocates argue that they are not interested in replicating the past in the same way that reintroduction projects do.[66] Many of the species targeted in rewilding efforts also have been gone from a given landscape for many hundreds or even thousands of years, which contrasts with standard reintroduction projects that focus on recent extinctions. Rewilding initiatives advocate the creation of more "wild" nature that is left to its own devices, meaning that there should be less human intervention in the landscape.

In chapter 3, I turn to a historical case of rewilding with muskoxen in Norway and Sweden from the 1930s. Muskoxen had been absent from the Scandinavian Peninsula for several thousand years before this modern rewilding project. For those involved in bringing the species back to Scandinavia from Greenland, the muskox symbolized hope. They thought of the animal as a potential contributor to a more productive landscape and operated within an optimistic outlook. Hope motivated their decisions and claims that muskoxen belonged in Scandinavia. Muskoxen were not greeted warmly by all the rural inhabitants, however, illuminating potential conflicts with rewilding. For those on the ground, fear was a dominant emotion.

Although the word *rewilding* is somewhat anachronistic for the early twentieth century because the word is no more than thirty years old, the practices centered on the muskox were remarkably similar to contemporary attempts to rewild Europe with long-gone megafauna. Projects like Oostvaardersplassen in the Netherlands, which has used Heck cattle, Konik ponies, and deer to recreate an open grassland environment; and Pleistocene Park in Russia, which is rewilding a huge area to recreate the "mammoth steppes" of Siberia, envision megafauna as potential saviors of the countryside.[67] Returning to a radically wilder landscape is motivated by a sense of loss in the current landscape.[68] Because of the similarity with those contemporary projects, the muskox rewilding of Scandinavia offers a valuable historical view into what is generally portrayed now as a brand-new idea.

The newest of the three practices is resurrecting species that have already become extinct. The resurrection of extinct species, often called *de-extinction*, can happen through manual back-breeding to activate specific traits characteristic of a specific extinct species or through genetic techniques to either clone a new living member of a species that has died out or manipulate genetic sequences of existing species to match the extinct species. Back-breeding to create "lost" species occurred in the twentieth century with the development of Heck cattle to replace the extinct aurochs and attempts to recreate the quagga, a zebra subspecies.[69]

Genetic-cloning techniques to resurrect extinct species are newer developments. For these kinds of projects, scientists attempt to sequence the DNA from the extinct species, then they map it onto existing related species. Scientists make genetic modifications to existing DNA used in embryo-cloning processes, and, if things go right, the embryo is implanted into a surrogate mother of a related species and results in a viable offspring with at least some of the traits of the extinct species.[70] There is rarely an instance in which *Jurassic Park* does not come up when de-extinction is discussed in either the press or scientific publications.[71] In *Jurassic Park*, a scientist successfully recovers dinosaur DNA and is able to genetically manipulate raw DNA material to recreate viable dinosaur embryos.[72] Of course, such a scheme goes terribly wrong and the dinosaurs run wild. Although scientists are always quick to point out the differences between current genetic work and the film's portrayal, noting that bringing back dinosaurs is not possible because DNA has not been recovered, bringing back wooly mammoths in this fashion is almost within reach.[73] Technically, we have already had one de-extincted animal: cloning techniques were used to recreate the Pyrenean ibex (*Capra pyrenaica pyrenaica*; also called the Spanish bucardo) in 2003 after the last living individual died in 2000, although the ibex fawn only lived for several minutes after birth.[74] Resurrecting species has already had limited success, and significant monetary investments in this area are likely to bring more, despite significant critique of the practice on ethical and philosophical grounds.[75] The use of biotechnology to address the extinction crisis may simply be the latest symptom of humanity's increasing reliance on technology to control nature.[76]

As I will discuss in chapter 4, projects to recover the passenger pigeon, Tasmanian thylacine, or mammoth attempt to find the lost through today's genetic technologies. By examining writings about the extinction of the passenger pigeon in 1914, both those nearly contemporaneous to the event and those one hundred years later, this chapter will examine the work of grief as an emotional longing in passenger pigeon history.[77] The passenger pigeon's extinction has been framed as a tragic loss, but the way grief manifests itself has changed over time, including anger, depression, and denial. Grieving for the passenger pigeon is now being translated into species-resurrection attempts, which received significant public attention in conjunction with the centenary anniversary of the passenger pigeon's extinction. The idea is that extinct species may yet return, which upends the depression phase of grief by returning to a hopeful bargaining phase. As Carrie Friese has remarked, "Reproducing hope is indeed a key part of the rhetorical work of de-extinction."[78]

In chapter 5, I reflect on the ways in which histories of the recovery of nature are remembered. Within the walls of the natural history museum and the unbounded fields of nature tourism, animal reintroduction, rewilding, and resurrection events have the potential to affect how the histories of specific species, as well as larger environmental histories, should be told. I will examine some places in which these stories have been integrated, as well as where they have not. Chapter 6 argues that emotions are powerful in the telling and retelling of recovery stories. How we tell histories like the ones in this book creates a history of the past for the present and future.

The emotional attachments to past environments and to acting to recover those things that are thought to belong but have been lost fundamentally shape modern conservation. The way that the past is remembered, how emotions are wrapped up with the stories, and how the past is mobilized in the recovery stories are critical. As Jan Assmann, a scholar of memory, has commented, "The past is not simply 'received' by the present. The present is 'haunted' by the past and the past is modeled, invented, reinvented, and reconstructed by the present."[79] This is why storytelling about the past matters. In this cultural environmental history, I am going to pay attention to the stories people have told about recovering nature—especially how human-nonhuman relations of the past are deployed as historical narrative, with particular emotional slants to justify species interventions.[80]

2 Reintroducing: Guilt and the Beaver's Return

The sun was still up even though it was after midnight. A crowd of twenty—nineteen men and one woman—stood around the Leipikvattnet lake in anticipation. Dr. Sven Arbman, a young zoologist who had accompanied the box that now sat ready on the lake shore during its four-day journey from Stockholm to the mountains of Jämtland in Central Sweden in July 1922, spoke to the crowd. According to Arbman, the two beavers that crouched in the box before him were part of a "revolution in which we participate." With the release of the beavers, part of the damage humans had done to nature would be reversed. Then Eric Festin, the director of the local cultural museum who had put the event together, said some words about a landscape reinhabited by beavers as "a future land." The old men's eyes teared up at the thought of a manifest past, present, and future, of the beaver being back where it belonged. It was half past three in the morning by the time they were ready for the culmination of two years of hard work. The beavers were lifted out of the box and set down beside the water. They scampered quickly into the water, swimming straight into the lake. The beavers smacked the surface with their tails. The camera flashes tried to capture the event, but it was hard to see the dark bodies flittering in the water even under the lit sky. By five in the morning, the party dispersed and the beavers had disappeared into the wild. After an absence of about fifty years, beavers were back in Sweden.[1]

It is not often that an event in a remote area of northern Sweden has lasting national and international significance, but this one did. This was the first time that European beavers had been released into the wild to recolonize their former habitats.[2] It would inspire others and provide a template for beaver releases throughout Sweden, Norway, Finland, and the Baltics, making beaver reintroduction the most widespread and successful animal reintroduction activity in history. The beaver's return in the interwar years was not about ecological restoration writ large, but rather about recovering one species that had been identified as lost. The people involved in the reintroduction might have appreciated ATM's street art; as I will discuss in this chapter, they too worked to find the lost beaver and return it to the landscape in which they thought it belonged.

Yearning for lost nature carries a strong impulse to lay the blame for its disappearance. Unlike ATM—whose poster which declares that the beaver is gone, perhaps run away of its own accord—the proponents of the Swedish reintroduction named humans as the reason for the loss of the beaver. In their tellings of this history of extinction—and then the return of this species—human culpability for the beavers' loss was often a linchpin in the narrative. In other words, humans were guilty and need to atone for past sins.

Guilt is typically understood as the feeling that results from an individual's knowledge that he or she acted against his or her own moral or ethical standards. Often the impulse is then to make recompense for the action: "When someone feels that he has done something wrong there will be a tendency for him to make up for his wrongful deed."[3] This applies to environmental actions, as well as personal relations: researchers have demonstrated that consumer behavior such as buying environmentally friendly products may be influenced by personal guilt.[4] Collective guilt also has been theorized as a reason for some environmental actions. In particular, research on climate change has argued that individuals who feel guilty that humans, including themselves, are the cause of the change are more willing to engage in mitigation.[5] Prior research into environmental action and guilt shows that people feel guilty for actions right now degrading the earth, such as recycling habits, new consumer purchases, and CO_2 emissions.

In this chapter, I will show that environmental recovery discourses can operate on another level: they create a sense of guilt over the actions of our forefathers in the distant past. The living people who worked to bring back the beaver had no hand in its extirpation, yet they *personally* felt guilty that the animal was gone and wanted to make amends. The actors *regretted* this extinction history. Although regret is often associated with feeling negatively about something you personally have done, this story reveals a form of corporate regret. The discourse of collective guilt and its concomitant regret makes a person responsible for not only personal actions but also the actions of all humankind who came before. What I will show here is that the loss of the beaver and longing for its return was framed by feelings of personal guilt over the beaver's extinction at the hands of others.

The Lost Beavers of Sweden

The European beaver, which has the scientific name *Castor fiber*, is the continent's largest native rodent, measuring 1–1.25 m

from tip to tail and weighing 20–25 kg as an adult. Its striking, oversized, orange-enameled incisor teeth are used to gnaw
branches and tree trunks. These teeth assist the beaver in eating
a wide range of vegetative materials, including aquatic plants,
fruits, leaves, twigs, and tree bark. Also memorable is the beaver's
large, flat, scaly tail, which is used as a rudder when swimming
and as a warning signal by slapping the water surface. Although
most people envision beavers as dam builders, the European beaver, in contrast to its North American cousin *Castor canadensis*,
rarely builds significant dams.[6] Instead, these beavers tend to
block streams with smaller structures to raise the water levels
and live in burrows on the sides of the streams. Beaver structures
affect water drainage patterns and increase the amount of wet-
land and riparian habitats.[7] Beavers live in monogamous pairs
and mate for life. They generally have two or three kits born in
the late spring. In addition, several juveniles under the age of
two often cohabit with their parents. A pair will tend to stay in
the same place unless something or someone makes a radical
change to the environment.

Castor fiber and humans have a long history of contact. The
animal once inhabited a large region from Western Europe to the
Chinese-Mongolian border.[8] A multitude of human pressures—
hunting for meat, fur, and oils; destruction of beaver canals and
small dams; and deforestation—caused the beaver's precipitous
decline in Europe by 1800, but the animal already may have
been quite rare even in the medieval period. When Gerald of
Wales described the beavers he saw on River Teivi in Wales, he
called them a "particularity" because they still survived only
there in Wales and were found in only one Scottish river.[9] There
was some medieval trade in beaver fur, particularly coming out
of Russia, but the beaver stock in Continental Europe appears

to have been too small to make fur trading profitable.[10] We do have evidence, however, that beaver pelts were a traded from local trappers in northern Sweden to continental ports such as Amsterdam in the 1700s.[11]

Despite dwindling numbers, some beavers in Europe were killed for food, as evidenced by recipes for beaver in medieval and early modern cookbooks. In his highly influential *Natural History*, Pliny the Elder classifies them as *amphibious animals*, a categorization that carried over to cooking manuals, such as John Russell's *Boke of Nurture* from circa 1460, which starts the "Carving of Fish" section with the recommendation to serve beaver tail as the meat in pea soup or frumenty (a kind of cracked wheat porridge).[12] A 1650 version of *De Alimentorum Facultatibus* (On foodstuffs) likewise starts its chapter "De Animalibus amphibiis" (The amphibian animals) with beaver, followed by otter and frog. The classification of beaver as an aquatic animal, instead of as a land animal, may have made it acceptable to eat beaver during Christian fasting seasons such as Lent. Gerald of Wales wrote that "in Germany and the arctic regions, where beavers abound, great and religious persons, in times of fasting, eat the tails of this fish-like animal, as having both the taste and colour of fish."[13] Beaver meat consumption, particularly by hunters who would then go on to sell the skins and scent glands, appears to have been common in eighteenth and early nineteenth century Sweden.[14]

Beavers in Sweden were hunted as much for medicine as they were a source of furs or meat. Beavers have scent glands called *castoreum sacs* located near the base of the tail. Humans harvested these glands to extract the castoreum, often by either drying them to make a powder or infusing the glands in an alcohol to cause the castoreum to leech into the liquid. The importance of castoreum as a motivator of beaver hunts is evident in an ancient legend stating that when beavers are being hunted, they stop and bite off their testicles, which are then thrown to the hunter so that the hunter receives what he is after and the beaver is allowed to live. The legend confuses the castoreum sacs with testicles, but it is understandable: the castoreum sacs are paired, so they resemble testicles; they are located at the base of tail near where testicles would be expected; and beaver testicles are not descended, which makes it appear that the male beavers have lost theirs. This legend, which is repeated from the works of Pliny the Elder to the early modern period, is commonly used in illustrations of the beaver in books of beasts (figure 2.1).

According to Pliny, the oily castoreum was used in a variety of medicinal applications, including being taken internally to treat brain swelling, epilepsy, bowel irritation, vertigo, and paralysis, among others, as well as being applied topically to reduce bite and sting inflammation, toothaches, and earaches.[15] Arabian medieval medicine employed castoreum in preparations to treat headache, urinary leakage, cramps, and mental illness. Such medicinal uses for castoreum were common until the twentieth century. The Norwegian pharmacopeia of 1879, for example, contains recipes for two castoreum tinctures.[16] Old apothecary collections in museums almost invariably have a jar of castoreum. Oral history accounts of events in the early nineteenth century measure the beaver catch by the weight of the castoreum glands and note the medicinal uses for the beaver. For example, when the family of Olof Jonsson started trapping beavers in Flakaträsk in northern Sweden in 1810, they reported catching eleven animals with a total of 150 lod (approximately 2 kg) of castoreum sacs, which were sold along with the skins in Umeå. In 1828, hunters caught one beaver that yielded castoreum glands weighing 92 lod (approximately 1.2 kg).[17]

Figure 2.1
Scene of a beaver castrating itself before hunters in a medieval bestiary. Bestiary, second quarter of 14th century, MS. Bodl. 764, fol. 14r. *Source:* Bodleian Libraries.

Beaver encounter stories told in the early twentieth century about the nineteenth century tend to focus on castoreum. Alarik Behm, the director of the Skansen zoological garden in Stockholm, told a tale focused on castoreum in his book *Nordiska Däggdjur*, published in 1922: "My grandmother, born in the

Jämtland mountains at the beginning of last century, used to tell us kids, that when someone in her home town was very ill and began shaking ..., you would give the sick person castoreum, a half teaspoon in a sip of liquor, after which the patient died. Either the medicine was given too late or else it had no effect. This castoreum was supplied by two men, Halvar and Marten, and thanks to their and others' earnest 'supplies,' beaver disappeared from the area."[18] Regardless of the effectiveness (or not) of the castoreum, it seems that medicinal use of beaver gall was one of the primary drivers in Sweden for hunting it.

Even in the remote Scandinavian forests, the beaver was rapidly disappearing by the modern era. Swedish men of letters writing from the eighteenth century noted with concern the decreasing beaver numbers. In 1756, the naturalist Nils Gissler, a student of Carl Linnaeus, expressed concern that the beaver was being overhunted in Sweden. Before "they never caught all of the pairs in each place, and never touched the young," he wrote, but now "they kill all they can," with the result that the numbers were diminishing.[19] By 1832, the author of a major book on Swedish hunting, G. Swederus, believed the beaver was "only in the northern part of the country and in the wild tracts" because the animal was being pushed there by growing colonization.[20] The major scientific treatise *Skandanavisk Fauna*, published in 1847, noted that even in the northern reaches, "extinction pressure has caused this animal to almost disappear."[21]

In 1873, Ferdinand Unander, the head of the agricultural school in Västerbotten County, published a review that declared the beaver extinct in Sweden. He had systematically reviewed known beaver sightings and found that the most recent evidence was from the far north in 1864, although the journal's editor added a footnote that another beaver was seen in Jämtland in 1866. Unander concluded "that as long as no proof is shown that beaver is found in the Swedish dominion and by which refute the before given facts and figures, he [beaver] must be regarded as an animal extinct from the Swedish fauna."[22] The once-numerous beaver was gone from the Swedish landscape.

The European beaver's decline was widespread and nearly complete. By the late nineteenth century, there were only scattered enclaves in Russia, Mongolia, Germany, France, and Norway; likely only about 1,200 European beavers remained in total.[23] The beaver's fate seemed sealed: to pass away in oblivion like so many other larger mammals under the human-dominated landscape.

Feeling Guilty about the Beaver's Extinction

Along with his declaration that the beaver in Sweden had become extinct, Unander suggested that the beaver needed to be brought back: "to make amends for the mistakes in hunting," he proposed reintroducing beavers caught abroad.[24] From this very first suggestion in 1873 to reintroduce beavers, guilt and regret were key motivators of the actions that would follow. The sentiment was that mistakes had been made in the past and needed to be corrected in the present.

When Erik Modin (1862–1953), a well-educated pastor from Jämtland with strong interests in cultural heritage and nature protection, asked the question "Shouldn't something be done to reintroduce beaver into our land?" in an article in a Swedish hunting journal in 1911, he wanted to make amends with the Swedish countryside.[25] He argued that the Swedes should return the beaver, "a pious and harmless" animal, as a "matter of honor and duty to seek peace and help it come home."[26] When

Modin framed the beaver as pious and harmless, he called on humans to be the same. He asked the readers to seek peace—to offer a truce to the beaver, which, although harmless, had been slaughtered. Modin posited an idea of corporate guilt for long-past sins of forefathers who had hunted the beaver to extinction. Although his readers were not personally guilty of eradicating the beaver, he wanted to instill a sense of personal responsibility for righting the wrong. In this, we see that Modin not only felt that he was guilty himself, but also believed others should feel guilty. For him, it was a moral obligation to return the beaver to its homeland, the place where it belonged.

Modin proposed reintroducing beavers to one of two northern Swedish national parks: Abisko or Sonfjället, which both were established in 1909. He believed that the national parks were appropriate locations for new populations of beavers because the animals previously had lived in the areas of the parks and would be protected there under national park laws. The national parks also were ideal because Modin believed that the park's mission "seems to me, to be not only to preserve and protect what the Swedish countryside has within it at the present time, but also to—in so far as it may be done without great difficulty—restore what once belonged to the same." The beaver's belonging was central to Modin's plea for action. Modin did not, however, act upon his suggestion. He ended his article by calling on others ("benevolent benefactors") to take up the task.[27] It would be another decade before someone would.

In July 1920, Alarik Behm (1871–1944), the superintendent of the zoological garden in Stockholm, stood on the podium at Sweden's second Cultural Fair (Kulturmässan) in Östersund. The Cultural Fair brought together a wide range of exhibits, from 1600s French art sent by the National Museum to regional church art to watercolors of Jämtland's birds painted by local artists.[28] Behm's talk, "Some Words on Nature Protection" (Några ord om naturskydd), was sponsored by the Jämtland and Härjedalen Nature Conservation Association. His lecture featured the beaver, which he claimed had been deliberately hunted down because it had been thought of as a forest devastator. He argued that beavers and people were not incompatible; beavers in Germany, France, and even Norway seemed to get along fine with the locals. He warned that other animals, like the red deer, brown bear, and lynx, might go extinct in Sweden too if changes in attitudes were not made.[29] Although Behm's speech didn't directly call for beavers to be reintroduced, the idea had been planted—and an appeal for funds for a reintroduction project would be made the next year by Eric Festin.

Eric Festin (1878–1945), educated in archeology and art history, had become the Jämtland County curator in 1919 and had organized the Kulturmässan event. He was a local boy, the son of a farmer in the village of Hackås, not far south of Östersund. An injury to his back as a youth had made him unsuitable for heavy labor work, so he was sent to recieve an education. He studied in Uppsala, and after graduating in 1912 with a bachelor's degree, he moved back to Jämtland. He became the county curator responsible for cultural and architectural heritage preservation in Jämtland, a position he would hold until his death. His secretary during his last five years characterized him as "a dynamic workaholic" who was in the office from 9:00 a.m. until 9:00 p.m., writing letters and reading. "I don't know if Eric Festin was a happy man," mused the secretary, "but if work and if finding the right job here in life is happiness, then yes he had that."[30]

Festin became secretary of the Jämtland and Härjedalen Society for Nature Conservation (Jämtlands och Härjedalens

naturskyddsförening) in 1916. This was the local branch of the
Swedish Society for Nature Conservation, the oldest environmental protection group in Sweden, founded in 1909. Although
Festin's work focused initially on cultural heritage, it was not
a big leap in early twentieth-century Sweden to bring nature
protection into the mix. In 1914, Karl-Erik Forsslund, a prolific
author and cultural heritage advocate, published a two-volume
work on cultural heritage protection, with one volume dedicated to nature protection and national parks.[31] According to
Forsslund, society needed to protect "wild nature" because it
was vital to the human soul: "Since we all have a soul, it is the
meaning that it shall live and grow, be healthy and well nour-
ished. But it must have nourishment and air in order to grow—
wide vistas in order for the soul to not be crowded and stuffy,
beautiful living sights so that it will not be ugly and parched."[32]
Culture and nature were profoundly linked. Behm's talk at the
Östersund festivities that inspired Festin to act had followed this
lead, bringing together cultural heritage and nature conservation: "Nature protection," Behm intoned, "is just one side of
cultural protection in a wide sense; yes, one can even say that
cultural protection is just one side of nature protection."[33]

The ties between cultural heritage and nature protection
played a large part in building interest around the beaver project. Jämtland, where the beavers would be released, had a history with beavers. According to Eric Festin, "the legends about
the beaver still live on on the people's lips."[34] These legends of
human-beaver interaction would be repeated time and again
in publications about the reintroduction efforts. Stories were
told: stories from old men from the Jämtland region of the great
beaver trappers and the slaying of the last beavers; stories from
grandmothers whose grandmothers used medicine made from

beavers; stories of how the beaver once lived and died on the land. The memories of the beaver, though the stuff of legend, were integral to the decision to reintroduce it.

The cultural history of the beaver was invoked again and again when discussing the idea of bringing the beaver back, to claim the beaver belonged in Sweden and that the people longed for its return. As Erik Modin wrote in 1911, the animal was "now known only by legends and stories of yore." He had brought some of those to life in 1907, when he published the transcription of a text written by Johan Winter of Flakaträsk, dated November 25, 1865, that described beaver hunting and extinction in the county of Västerbotten in the first half of the 1800s.[35] The text is filled with personal remembrances, as well as numerous oral histories of where beaver lived and how they were caught. In his 1911 article, Modin recounted the sighting of the last living wild beaver in Sweden in 1866 in the Juveln lake in Jämtland as an act of memory. Festin likewise would repeat human-beaver interaction stories in his publications, noting that Jämtland was one of the last places beaver survived, even though there was "an intense war of extermination" against the animal, which came to its ultimate conclusion when a professional hunter named Per Persson from Långsele killed the last beavers in Bjurälvdalen in 1850.[36] Festin stressed that beaver had "an ancient history in our country" and should be brought back not only because beavers once lived in Sweden, but also because of the cultural milieu.[37]

Festin placed the new beavers into an old memory landscape. Festin's writing was filled with folk stories about the beaver, justifying its return. Stories about the beaver had continued to be told long after the animal was extirpated from Sweden. According to Festin, old people he talked with knew stories about the beaver: "From the old people, I have heard a good deal about the beaver, but with only few exceptions, the stories have been dated to the narrator's grandfather or greatfather's experiences, often in even older generations' time."[38] With a background in architectural preservation and work as the county museum curator, Festin was particularly interested in discussing the place of beavers in traditional folk practices. For example, Festin began one article on the reintroduction by retelling the story about castoreum from Alaric Behm's book *Nordiska däggdjur*.[39] He also wrote about hunting practices that generated legends about "the ingenious tricky beaver," legends that "give evidence also of rich flourishing superstition, which certainly in many cases incited the hunter's passion."[40] For Festin, the beaver was not an animal in the countryside away from man, but rather part of the cultural traditions of the Swedish people. Festin was a cultural heritage specialist: as the curator of the Jamtli county museum, he was involved in architectural preservation and restoration, artifact collection, and oral history collection. For Festin, the beaver in the wild was no different from the historical buildings in the museum he was working to preserve.

All this discursive work claimed that the beaver belonged in Sweden and provides evidence that Modin, Behm, and Festin longed for its return. They had emotional attachments to the beaver and through their writings wanted to cultivate those emotions in others. The beaver was something that was missed in Sweden, and missed locally in Jämtland.

At the meeting of the local Jämtland and Härjedalen Society for Nature Conservation on March 29, 1921, beavers came up in the context of a discussion of an area in the county known as Bjurälvdalen.[41] An investigation of Bjurälvdalen—the name literally translates to "beaver river valley"—by Fredrik Svenonius in 1880 had shown that the area was dominated by a unique karst landscape with remnants of beaver inhabitation, including beaver canals and dams.[42] The naming of the valley was a strong part of the remembering, as it consistently invoked an element missing from the current landscape. Bjurälvdalen was an area that the Society for Nature Conservation was keen to protect, and linking its protection to beavers seemed like the perfect way forward: "As a geologic peculiarity of the first order Bjurälvdalen should be protected, and that at the same time Sweden's first resurrected beaver colony should have its sanctuary in this previous beaver paradise."[43] The religious imagery of a resurrection in paradise reflects well the nearly religious fervor with which Festin would pursue the reintroduction idea. It may also indicate that Festin thought of the reintroduction as redemptive: just as the resurrection of Christ had washed away the sins of believers, so would returning the beaver wipe away the guilt of its extinction in Sweden.

In the 1921 yearbook of the Swedish Society for Nature Conservation, Festin appealed to committed nature lovers across the country for contributions to pay for the beaver's return. Although the local Jämtland and Härjedalen chapter would lead the effort, it lacked the financial wherewithal to get it done. The beaver, Festin stressed, had a long cultural history in Sweden, being mentioned in provincial laws from the 1200s. People in northern Sweden still told tales of beavers, so it was a place that the beaver belonged. Yet the reintroduction was more than a local initiative, Festin argued. Instead, he wrote that it was a *"national measure for all"* that "would certainly, not least among the young generation, make an impression and create a broader and better understanding of conservation efforts."[44] There is a nationalist angle in Festin's appeal, one that previously appeared in Behm's statements contrasting the beaver's survival in other countries. The beaver reintroduction was seen as something that would be good for Swedish nature, which had previously been harmed by human action.

Finding the Lost Beavers

Although beaver had become extinct in Sweden, they had not become extinct in the neighboring country of Norway—or at least not completely. Like elsewhere, the European beaver had suffered significant declining populations in Norway. Studies in the late 1880s by Robert Collett (1842–1913), a professor of zoology with the University of Oslo Zoological Museum, paint a picture of beaver as species struggling in Norway. He had done extensive fieldwork and found the beaver existing in significant numbers in only two places in the country: (1) the Nid river in the south, particularly near Aamli; and (2) the Tørenæs river above Drangedal. Both places are located within 100 km of each other in the eastern part of the southern peninsula of Norway. He estimated only one hundred individuals living in the Nid watershed and even less on the Tørenæs.[45] Hunting of beaver had been regulated in Norway under the Hunting Law of 1863, which allowed beavers to be killed only in August, September, and October by the landowner himself, but even at the end of the century, Collett thought the beaver population was not increasing significantly.[46]

Yet official governmental protection, as well as informal protection by particular landowners who banned killing beavers altogether, provided enough space for the Norwegian beaver numbers to rebound by the 1910s.[47] In 1918, an open hunting season of fourteen days in the fall was instituted because of the growing number of forestry damage claims attributed to beavers.

Sigvald Salvesen, who supplied beaver specimens from the Aamli area to numerous museums, estimated that the meager population of one hundred beavers along the Nid river in 1883 had grown to a whopping twelve thousand by 1928.[48]

So when Eric Festin thought about bringing beavers to Sweden, he turned to Norway. Festin commissioned Sven Arbman, a biology teacher in the town of Sunne, to determine how to acquire beavers in Norway.[49] In late August 1921, Arbman went to Åmli, where the Norwegian beaver population lived in healthy numbers. There he met with several potential suppliers, including Peder Martinus Jensen.[50]

P. M. Jensen (1881–1963), a taxidermist, was already known in Swedish zoological circles. His early letterhead declared his expertise as a taxidermist, with drawings of taxidermied specimens and images of medals he received for taxidermy work overlapping the pictures. The tagline sold Jensen as "provider of taxidermy for all sorts of birds, fish, animals, and animal heads. Living beavers (Castor fiber) for reintroduction and for zoological gardens." Einar Lönnberg, who was a professor of vertebrates at the Swedish Natural History Museum (Naturhistoriska riksmuseet), had used Jensen to supply some beaver specimens for his scientific collection.[51] In early fall of 1921, Behm recommended to Festin that he should contact Jensen to acquire beavers for reintroduction.[52]

"Beaver-Jensen," as he would affectionately come to be known, turned into *the* Norwegian beaver expert. He captured the wild beavers, took care of them in captivity, built specially designed transportation crates for them, and on more than one occasion traveled with the beavers to international reintroduction sites. By the 1930s, he primarily used a letterhead with three images at the top, all showing Jensen caring for the animals, including one photograph in which he gently cradles a baby beaver feeding on a bottle (figure 2.2). Jensen often was depicted in newspaper coverage in similar positions, seemingly cuddling the beaver. According to a close friend, "he and the newly captured beavers understood each other."[53] The nurturing image of Jensen is supported by others who likewise characterized him as "a beaver enthusiast who believes beavers are the best of all the animals."[54]

After Arbman visited Åmli in August 1921, Jensen was eager to be selected as the beaver supplier. He wrote a follow-up letter to Festin, stressing his prior experience with Swedes (naming Lönnberg) and offering a pair of beavers for NOK (Norwegian kroner) 1,000 to be delivered that autumn. The practicalities of repopulating the Swedish countryside with beavers were beginning to fall into place, but Festin still needed to pay for the whole operation.

Figure 2.2
The letterhead of P. M. Jensen, showing him taking care of his beavers. Letter in Jamtli Archive, Östersund, Sweden.

Festin estimated that in total they would need SEK (Swedish kroner) 3,000 to pay for the beavers, their transport, and the costs incurred by Arbman. He made his public appeal for funds in the pages of the Swedish Nature Conservation Association 1921 yearbook. In May 1921, one big donor had already come forward: Gustaf Werner (1859–1948), a textiles industrialist from Göteborg in southern Sweden, donated SEK 1,500 to the cause, "to bring again this interesting animal into our fauna."[55] But it wasn't just members of the upper echelons of society who supported the beaver project. The list of donations to Festin's "beaver fund" includes many very small donations: nearly all are under SEK 5, and many are only SEK 0.25.[56] The story of an encounter on a train during the beaver transfer exemplifies the way funds were raised for the project:

> A group of farmers, on their way home after a forestry course, dis-
> covered after the lecture that they forgot to pay admission and so
> we "scrounged" together 20 kr into our beaver fund. Superintendent
> Festin in Östersund decided that beaver would be set out in Sweden,
> and then it must surely happen. With a fund there is surely some
> means. "Is it worthwhile trying," asked one man here on the train,
> "since you can not guard the treasure up in a mountain valley?" "We
> will rely on upland dwellers," I said. "So it is," he said and gave a coin.
> From one poor village in Frostviken we have taken in 84 kr. And been
> promised more. And here a promise means something.[57]

Bringing back beavers, a "treasure" of the Swedish fauna, was a people's project. The donors came from a wide range of social classes, and so did the men who helped with the practicalities. Whether these financial donors had the same sense of guilt over the beaver's extinction as the organizers is impossible to know. But we do know that those who spearheaded the effort time and time again referred to their guilt and the reintroduction as amending a wrong. The beaver had been lost but would be found.

Welcoming the Beaver Home

A pair of beavers spent the winter of 1921–1922 in Stockholm at the Skansen zoo. They had been shipped by Jensen via train and then kept at the zoo until they could be taken to the northern Swedish release site in the summer of 1922.[58] At the end of June, the beavers were ready to depart. They were put back into the same wooden crate they had been sent in from Norway, and a four-day-long trek began. Arbman and Festin picked up the beavers in their box from Skansen on July 2 and set off by train to Strömsund, as far as the line would go. The journey from there grew significantly more complicated, requiring numerous transfers between horse-drawn wagons, sleds, and boats to pass through the mountains, valleys, and large lakes. At one point toward the end, the box had to be carried by hand like a coffin with pallbearers for three kilometers. The beavers were tossed this way and that in the box, even though Festin tried to cushion the ride by setting the box onto a base of hay instead of directly into the wagons. One of the beavers developed a fever on the third day, which Festin blamed on the jostling within the box. Luckily for the project, the fever passed.[59]

Along the road, the beavers and the story of the reintroduction attempt were shared with the local population. Each stop to switch transportation modes generated much public interest. Arbman gave natural history talks about the beaver at each stop and, according to Festin, urged the children to have the first look at the beavers, saying "this is for the future."[60] These talks were a way of transmitting the memory of the event to the spectators, particularly the children, who were singled out as important recipients of the experience and narrative. Storytelling is about transmitting memory to the next generation, as well as transmitting it to new places. It's the way that memory travels through space and time. The beaver's return was a memory-making event, with speeches and photograph opportunities.

Over and above sharing the memory of the event with those present, Festin thought it was important to document the beaver's return for posterity. He hired Nils Thomasson, a professional photographer from the area, to document the trip from Strömsund to the release point.[61] Thomasson's series of photographs exists in the Jamtli archive. Most of the photos focus on the transport mechanisms of the voyage with the beaver box rather than the beavers: the box is shown on a horse cart, being carried on poles, in a boat, and being pulled on a horse sled. There is only one extant picture of the beavers themselves in the Jamtli archive, although others may have previously existed: the beavers are being examined by Arbman, who was responsible for their health during the journey (figure 2.3).[62] In Festin's article about the release, the story builds to the climax as the beavers are lifted out of the box and set into the canal, "our cameras snapping rapidly, but pictures were difficult to take."[63] And indeed, in the series of photos, none of them show the beaver being released. Thomasson was on hand to pictorially record another beaver reintroduction in Jämtland in 1934, but it was Festin himself who managed to take a photo of a beaver in the water that time.[64]

Festin had also tried to get the beaver release captured as a moving picture. On June 23, 1922, Festin wrote a letter to Svensk Filmindustri, which was in charge of making all official Swedish films at that time. He asked the company to make a film about

Figure 2.3
Sven Arbman checks the condition of the two beavers to be released in 1922, while Eric Festin and a young boy look on. *Source:* Nils Thomasson/Jamtlis fotosamlingar.

the beaver reintroduction in Bjuräldalen, which would occur two weeks later. The reply he received read: "Unfortunately we cannot send a photographer to follow the expedition, which in any case we do not think from a film standpoint would be significantly rewarding. We already have educational films about the beaver and his lifestyle. This would be only the release and one or another expedition moments, which would not be worth the cost. We will happily shoot the beaver and expedition's start from Skansen, if you kindly inform us of it."[65] The existing educational films about beavers that Svensk Filmindustri had on hand were not, of course, about beavers in Sweden but on other beavers, but this nuance did not matter to the company.[66] It appears that no film was made, not even from the expedition's start—which is highly unfortunate because this first beaver project would inspire many others and was a watershed historical event.

Although the photographs rarely show the beavers, they do show people. There were twenty witnesses to the release of the new Swedish beavers. There were the main players—Festin, who led the event; Arbman, who cared for the beavers; and Thomasson, who took the photographs—but there were also many locals from the small countryside villages and farms of Leipikvattnet, Leipikmon, and Ankarvattnet. These were the individuals who had made their own transport modes available: horses, carts, and boats, as well as their shoulders at times to carry the box. All were males, except one young woman in attendance: fifteen-year-old Emma Karolina Jonasson, who was there along with her brothers.[67]

We do not know exactly what Arbman and Festin said at the reintroduction event on July 4, 1922, although we know the tenor of both speakers, based on remembrances recorded afterward. The event was framed as a cultural and natural restitution, entangled with God and country. Just as Festin had hinted that the resurrected beavers would be returned to a paradise on earth, Arbman alluded to a restructuring of the oft-cited command in the Bible's Genesis 1:28 that humans should "be fruitful, and multiply, and replenish the earth, and subdue it: and have dominion over the fish of the sea, and over the fowl of the air, and over every living thing that moveth upon the earth." To Arbman, this passage encapsulated the "order of nature" and the "march of progress," yet these were being overturned in the beaver reintroduction. It would no longer be the case "that mankind goes forth and multiplies and takes possession of the earth and self-evidently eradicates all other creatures if they are not good enough as man's slaves."[68] This echoed a passage on the last page of Forsslund's *Hembygdsvård* volume on nature protection: "We are the lords of creation, so we must be good and gentle lords. Mother Earth's all other children are our subjects, let us treat them not as prisoners and slaves, but as friends and helpers. We have a lot to thank them for, we should not pay them back by extraction and pillage, but revere and cherish them and their power and beauty."[69]

In building on Forsslund's view, Arbman believed the beaver reintroduction was changing the relationship between mankind and nature itself. He implied that humans had previously considered animals as their servants, even slaves, but humans were no longer entitled to exterminate animals they found nonuseable. He framed the beaver reintroduction as a new kind of progress. This progress made mankind not completely dominant over nature. In the reintroduction, the restorationists were serving nature, using their hands to release beavers to their former homes, rather than asking nature to be subservient. Festin also understood that these conservation actions were rearranging relationships, encouraging further reintroductions to "make up for one of the worst sins that ever happened to our Swedish fauna. *Give the wilderness what belongs to the wilderness!*"[70] The project was both recompense and restitution for past sins. Although the participants in the project had not committed the sin of extinction, they were still guilty of the crime.

Atonement through the Beaver's Return

By 1928, Festin could write of "the beaver's Swedish renaissance"[71] because of the number of beaver reintroduction efforts that had quickly copied the Jämtland effort. Between 1922 and 1927, forty-seven beavers had been imported from Jensen in Norway, although five of them had died en route—a reminder of the complications and strenuous effort required to move the beavers through multiple steps from southern Norway to their final release sites in middle and northern Sweden.[72]

The first group to copy the Jämtland reintroduction model was Västerbottens läns jaktvårdsförening (VLJ), the hunting association of Västerbotten County. VLJ was founded in 1919 by "interested hunters and friends of nature."[73] According to the first director of the organization, *jaktvård* (hunting stewardship) was considered something akin to household management that required care and limits; hunting laws were necessary to protect animals from widespread *jaktlust* (hunting desire). The director lamented that many people either did not respect the hunting laws or tried to kill as many animals as possible once the hunting season started. The association's task thus was to arouse interest in hunting "based on love for the animal world and nature"—a love that leads a hunter to "rejoice" when he sees "his wildlife thrive and multiply."[74] This was not the same kind of sportsman code that came into circulation in North America, which stressed hunting as a sport and diversion for modern men, as discussed by Tina Loo in *States of Nature*.[75] Tom Dunlap has argued that American hunting at about the same time was neither about conservation (sustainable management of the populations) nor preservation (completely protecting the animals) but "rather [was] a way to recapture the past and its virtues."[76] In his telling, hunting was a ritual activity that demonstrated mastery of nature and manliness.[77] Dunlap admits that hunters and hunting groups were effective at restricting hunting seasons and techniques in the United States during the late nineteenth and early twentieth centuries, but "the desire to hunt supported the impulse to save and to protect."[78] In northern Sweden, there does not appear to have been the same distinction between sport and subsistence, nor was there an emphasis on hunting as a modernizing project, as a way of getting back to nature. Man had never left nature, according to association members; the call was for greater love and respect for the nature in which man still found himself. Yet there was also a desire to recapture the past—a past that included beavers as an integral part of the Swedish countryside.

It was in this vein that the association became involved in beaver reintroduction. Rather than touting human mastery over nature, the association advocated stewardship of it, just as Arbman and Forsslund had. When Axel Sylvén, the VLJ vice chairman, proposed that the association reintroduce beavers in Västerbotten, he invoked the long history of the beaver in the country: it was "the beaver's true abode." The beaver belonged in Västerbotten, yet humans were guilty of exterminating it, according to Sylvén. He argued that the national government had failed to protect beavers by waiting too long to ban hunting and said that the association needed to work to rectify the loss.[79] Unlike in North America, where game extinction was blamed on commercial exploitation or indigenous peoples, the rural populace that hunted beavers and the central government that failed to control the hunt were considered equally to blame in Sweden.

A monetary fund was established and earmarked as the *Bäverfonden* by VLJ. The group anticipated that it would need SEK 3,000 to 4,000 for the effort and received a generous SEK 1,000 donation from a merchant, Ando Vikström from Sundsvall. Sylvén urged others to donate to the "patriotic" cause to bring beavers back to their homes that had stood empty for a century; the beaver's was "a sad fate" that served as a reminder "of where man's foolishness and rapacity can lead."[80] This was clearly a case of corporate guilt that needed atonement.

VLJ soon raised enough money to purchase some beavers from Bäver-Jensen. Unfortunately, the first three beavers VLJ purchased in 1923 died while overwintering at the Skansen zoo. According to Sylvén, zoo visitors brought about the untimely death of the first beaver: "The good Stockholmers could of course not resist cramming into our beaver all conceivable food with the unfortunate result that the beaver had a severe intestinal illness, which ended with its death in the spring of 1923, just when we hoped to acquire one or more comrades for release in Tärna during June or July."[81] Then, in September, the association purchased two replacement beavers that were sent to Skansen, but just after New Year's Day in 1924, the association received word that these two had also died. The beavers had contracted a contagious disease carried by a hare imported to the zoo from Finland. VLJ appeared to be having a bout of bad luck, but it was undeterred. The group bought another beaver in 1924, which was held at Skansen before being transported north to the release site. Four more beavers were shipped from southern Norway directly to Mo i Rana in northern Norway by boat, then traveled by cart across the Norwegian-Swedish border to Tärnaån, which allowed them to bypass the potential deathtrap in Stockholm.[82]

The beaver's homecoming was an act to make amends. Forester (*jägmästare*) Gunnar Esséen, state official (*länsfiskal*) Åström, and the association's treasurer and secretary in Umeå, Lennart Wahlberg, spearheaded the group of sixteen that trekked up into the remote mountains of Västerbotten for the reintroduction effort. At the release, Åström discussed the undertaking as landscape restitution: the "beautiful" mountain—"God's free nature"—had become "richer in immeasurable degree" through the "commendable deed."[83] Returning the beaver that had been lost made up for the errors of the past.

In texts published in the association's yearbook, the relationship between the hunters and beavers is not described in terms of conquest or domination, but rather co-citizenship. Both human and beaver belonged in Sweden. One photograph shows a member of the reintroduction group with a beaver, captioned "State forester Johansson cheering up one of his new 'countrymen.'"[84] A long article by Lennart Wahlberg begins by listing the "baptism officiant" (*dopförrättare*) and "godparents" or "sponsors" (*faddrar*) for the reintroduced beavers.[85] The use of these titles signals that these were more than people who observed the reintroduction; they were the people who became responsible for the beaver's success. Just as a baptismal sponsor agrees to raise a child to know God and the church, these individuals were agreeing to ensure the integration of beavers into the landscape. The beaver was an integral part of Sweden.

The interest in beavers was not just a passing fancy for the association. As G. H. van Post observed in 1930, "Surely all Västerbotten residents, and in particular the readers of this journal, are very interested in getting news about the beaver people's work and life in that wonderful beaver country."[86] The association maintained the "beaver fund" on its account books to pay for keeping an eye on its beavers, and a "Bäverrapport" was included in every yearbook through 1935. These reports noted where beaver tracks had been spotted, trees downed, lodges constructed, and young observed. The reports noted great anxiety when the beavers could not be located or appeared to have abandoned a given area, and great relief when they were seen again.[87]

In the VLJ publications, the hunting association members attested to having a duty to right the wrongs of the past, which had not hitherto been unaddressed by the central government. They did not have in mind *preservation*, in which nature is left to its own, or *conservation* of existing nature for future use—but neither was their motivation the "desire to hunt," as identified by Dunlap. Instead, these hunters felt they had to nurture nature as a godfather, to watch over the reintroduced beavers and help them succeed in their new-old homes as a way of expunging their own corporate guilt over the beaver's extinction.

Losing and Finding

The reintroduction of beaver in Europe is now hailed as one of the great conservation success stories of our time.[88] Reintroduction efforts continued, and by 1940 beavers had been released at nineteen locations in Sweden.[89] Although there was a pause for World War II and recovery time afterward, from the 1950s on beavers were caught in growing colonies within Sweden and relocated to other streams to aid their colonization.[90] The beavers rapidly spread across the landscape, and by the 1990s the population likely exceeded one hundred thousand individuals.[91] The beaver had returned.

The beaver was not reintroduced into Sweden—or other countries in first half of the twentieth century—based on its ecological function or as a biodiversity protection measure as we would understand it now. Rather, some years after becoming extinct, a group of nature lovers began framing the animal as something lost. Swedish nature had a gaping hole in it—but so, it seems, did culture. Nostalgic feelings for the time when beavers had been hunted, eaten, and used in medicine bubbled to the surface in the writings from those involved in the reintroductions. The idea was that the beaver belonged to Sweden and needed to be returned there. Finding the beaver in the correct place again was redemptive, a washing away of the sins of ancestors long dead.

This was, of course, a one-sided remembering. Very rarely did the negatives of beavers, such as downing trees in gardens or drowning farmland with water backed up from their dams, appear. When the forestry director of the Norwegian Ministry of Agriculture, Henrik Ielstrup, heard that a fellow forester in Orsa, Sweden, was considering reintroducing beavers, Ielstrup warned his colleague that he had received complaints about beavers cutting down fruit trees and damaging gardens. Yet even with that experience, Ielstrup did not want to sound like he was against the reintroduction efforts: "Of course, I am in agreement with you that from a nature protection standpoint it is interesting to preserve some beavers, so that it is not totally extinct in the Nordic fauna—but everything in moderation. I think that the interest in setting out beavers now should not be extended to a very large extent. But of course I would be happy to be helpful if you should want to buy beavers for reintroduction in your area."[92] In the available documents, Ielstrup's appears as a lone cautionary voice.[93] All the others seem to take it for granted that the beaver belonged back *everywhere* in Sweden and Norway. People like Festin were much more concerned about keeping people away from the beavers to protect the animals than the other way around.

As shown in this chapter, a profound sense of loss operated on the psyche of the Swedes involved in the beaver reintroduction efforts. They felt guilty about the extermination of the beaver, which they profoundly believed belonged in Scandinavia, even though they personally had not been the ones who killed off the beavers. Much effort was expended to get beavers again in Sweden—time, money, and animal lives had all been spent.

When Erik Geete, a forester, reflected in 1929 on the history and success of the early beaver-reintroduction efforts, he argued that the effort to return the beavers had been well worth it: "After the old population of beavers became extinct in Sweden, and our fauna become a species poorer, it took a long time before we realized what this loss meant. One can certainly give large thanks to the newly awakened conservation movement and their leaders for the thinking which brought about a revival of the beaver population. These men, who in this way spearheaded the efforts to give back this valuable and interesting animal *to Swedish nature*, are without any doubt worthy of our appreciation and gratitude."[94] This animal, about whom stories were still told and of whom place names still bore witness, had captured the hearts and minds of nature lovers in early twentieth century Scandinavia. Somehow the thought of a Sweden without beavers was unbearable; nature was not nature without beavers. Because man was guilty of its extinction, man had to make amends. The beaver, as much for cultural reasons as natural ones, was welcomed back to Sweden.

This kind of corporate guilt for past environmental damage continues to weigh on the minds of conservationists and scholars alike. Consider, for example, how emeritus professor of history Dan Flores, who specializes in the environmental history of the American West, frames his study of the loss of megafauna on the Great Plains: "Audubon was disgusted, 'even now there is a perceptible difference in the size of the herd, and before many years the Buffalo, like the Great Auk, will have disappeared.' He continued, 'Surely this should not be permitted.' But, to let Audubon's passive construction stand, it *was* permitted. *We* permitted it. We stood by and allowed what happened to the Great Plains a century ago, the destruction of one of the ecological

wonders of the world."[95] Collective historical guilt operates for
Flores as the motivation for a return to the wild plains he proposes later in the book. Now, we as readers can ask: Who is the
we in this construction? It is a *we* from a century ago—yet also
the *we* living today, we who never ourselves hunted the buffalo
or the pronghorn antelope that once swept across the landscape.
The *we* appears to be universal, but how might background, race,
gender, or economic status affect who is included in the indict-
ment? Flores doesn't say. His use of *we* functions to assign corporate guilt for human actions in the long-distant past. There is a
longing for that which belongs but has been lost because of us,
whoever that *us* may be.

3 Rewilding: Hope in Muskoxen and Fear of Wilder Ways

In October 1940, the Norwegian national newspaper, *Aftenposten*, ran a feature article with the headline "Dovre Mountains' newest ornament—Muskoxen doing well. There is room for thousands of them."[1] The article was well illustrated, with a muskox head to the left of the headline, five photographs of muskoxen in the snowy Norwegian mountains, and one image of a calf onboard a ship. The story of the muskox in Norway was told only in brief: ten animals from Greenland had been "planted out" in the mountains in 1932, and after some challenges to the population, the numbers were climbing. Although the headline made a point that the animals were "new," they were also presented as a species with a history of being wild and free in Norway: "Muskoxen are doing well in our high mountains—they have it so immeasurably free and peaceful in the wide open spaces up there. The species is old enough in Norway. In olden times they were found in Europe, and here they were found in the Dovre area—specifically during the construction of the Dovre railway line, bones of a muskox were found." The images reinforced the muskox as an integrated aesthetic part of the Dovre mountain landscape. The animals, which could multiply into the thousands according to the wildlife manager interviewed for the article, Jon Angaard, were portrayed as a welcome restoration of the mountains. There was hope that they would in some sense complete the landscape, which was deemed deficient of wild animals.

Yet there was in the article a recognition of the potential danger of living near muskoxen. In the opening paragraph, the writer points out that though muskoxen attacks are not common, when meeting a muskox "one should never feel too certain that it won't attack." He had had his own muskox close encounter in Greenland, and although he was not injured, he had felt the fear of facing a charging bull. Angaard downplayed the dangers ("I can't say they are dangerous," he said) but did admit that a muskox would attack if people got too close.

The tension between the hope that muskoxen will be a wild, natural part of the mountainscape and the fear that muskoxen will be a wild, dangerous part of the mountainscape arises again and again in this history. It's a story of conflicts encountered in making the Norwegian mountains wild again through the return of a lost species that had been missing for thousands of years—a practice we now call *rewilding*.

Rewilding as an Environmental Paradigm

Although *rewilding* as a word doesn't appear in print until 1998, I use the term to describe a concept that has historical applicability too.[2] The tale of muskoxen in Scandinavia resonates with modern rewilding efforts in several ways. First, it's the reintroduction of a megafauna species that had been absent from the area for hundreds or even thousands of years. This has been a common thread of rewilding efforts, which have tended to focus on large mammals such as the European bison in Eastern Europe (gone from most of Europe by the eighteenth century, although the last free-range wild animals didn't die until 1927), the Eurasian brown bear in the Alps (missing from most of its central and southern European range for about two hundred years), and beavers in Scotland and elsewhere in the United Kingdom (extinct for about five hundred years there). This focus is different from that of commonplace reintroduction projects, which focus on recently extinct animals. Second, tourism comes into play as one of the key economic reasons to keep muskoxen in Scandinavia, and this is also a major argument for twenty-first-century rewilding. Rewilding Europe, an international conglomerate of rewilding initiatives, profiles nature tourism as a way to make rewilding create economic benefit for the local community.[3] Typical reintroduction projects, on the other hand, rarely make tourism arguments as their justification. Finally, like modern rewilding, the muskox project built on the idea that wildness, rather than control or order, is a desirable outcome for nature management.

There is a palpable turn-back-the-clock flavor to much of the rewilding literature. Pleistocene rewilding, as proposed by Josh Donlan and colleagues in 2005, is a way of recreating the ecosystem of megafauna that existed on the Great Plains of the United States when humans first crossed to North America. The premise is that the extinction of large mammals in the Americas around thirteen thousand years ago created an "extraordinary impoverishment" through the "loss of their ecological roles."[4] For Donlan and coauthors, Pleistocene rewilding uses "ecological history as a guide to actively restore ecological and evolutionary processes rather than merely managing extinction."[5] In Siberia, Sergei Zimov is attempting to restore the "mammoth steppe" ecosystem of the tundra through reintroduction of large grazers such as European bison, yaks, and muskox. Although he doesn't yet have real mammoths to release, if the current efforts at resurrecting the mammoth (or a creature that is similar to it) succeed, he may have them soon. These Pleistocene rewilding efforts to return nature to its presumed former glory operate within a nostalgic paradigm. As Eric Higgs has observed, "Past landscapes, like old buildings, derive aspects of their value from nostalgia, continuity, and depth. ... Implicit in the act of restoration is the belief that such places are better than what now exists and worth bringing back. For many, restoration reflects nostalgia in the truest sense of the word: a bittersweet longing for something lost."[6] There is a yearning for a time before human interference behind the ideas of both Donlan and Zimov.

Journalist George Monbiot has become the most vocal international supporter of rewilding after the publication of his book *Feral: Searching for Enchantment on the Frontiers of Rewilding* (2013).[7] His focus is on Britain and turning what he sees as unproductive and harmful agricultural production land into space for wildlife, particularly wildlife that previously has been eradicated from the British Isles. The subtitle of the book is the most telling aspect: Monbiot is searching for enchantment. He argues that the changes wrought to the landscape and fauna make him "ecologically bored."[8] He wants once again to be exhilarated by and in the landscape. Like a fairy tale that begins "long ago in a kingdom far away," his quest for enchantment rests on a premise that in the past, things were better—and since we want better, we want the past. That past is filled with emotion: hope.

Hope was understood as a fundamental and positive human emotion throughout the medieval and early modern periods—for example, in work by Thomas Aquinas and David Hume—although it has been disparaged by some post-Kantian commentators like Friedrich Nietzsche, who rejected it as misguided faith in externalities.[9] Hope is an emotion related to an anticipation of something desired.[10] Importantly, hope is not simply a type of optimism; the person anticipating the desired result may not actually expect that it will come to pass, yet there is the possibility that it will.[11] The object of hope may be a concrete thing, or it may be a state of affairs. Hope involves evaluation because the person has to decide that the object of hope is desirable and worth getting.[12] Hopeful states "characteristically exhibit desiring, and believing possible, and sometimes they involve imagining."[13] This means that the temporal character of hope is future-oriented, yet the feeling of hope itself occurs in the present.[14] Because hope is focused on the future, it has the potential to be transgressive, urging the creation of conditions outside of the established status quo.[15]

Hope has the potential to move people to act in environmentally positive ways.[16] Geographer Laura Smith has argued that ecological restoration moves beyond narratives of guilt and restitution to redemption and hope.[17] For her, the redemption action is not a redemption of nature itself but a redemption of humanity: as humans, we can make ourselves right by repairing our relationship with nature through restoration, and that provides hope. The redemption is then a future hoped-for state, one that can be imagined and anticipated, without expecting with certainty that it will come to be.

As I'll discuss in this chapter, the people who brought the muskoxen to Norway were indeed hopeful—hopeful that muskoxen would repopulate the wild, hopeful that eventually they might be useful as meat or for wool, hopeful that the muskox's presence in Norway would atone for the large-scale slaughter of the animals in Greenland. Hope for restoring the lost wild can be a strong motivator in recovery action. Yet this chapter will also demonstrate that not everyone responds to the wild in the same way.

Although the instigators of the muskox rewilding actions expressed hope, those who lived in the area and faced potential dangers expressed fear. Fear as an emotion has both individual and social dimensions. At an individual level, fear of immediate threat has a physical effect on the body, but individuals also can experience a more vague fear of a future threat. When responding to potential futures, fear is the mirror image of hope: both can be based on things that have not yet come to pass but can be imagined. Fear is also socially constructed, and in the modern age, one of the fundamental fears in modern society is fear of an unnatural death.[18] Michael Egan has analyzed a growing awareness of environmental contamination and toxics in the twentieth century as part of this modern construction of fear.[19] In the Scandinavian case, death by muskox was understood as an unnatural death from an environmental danger. Those fears of muskoxen also are related to feelings of powerlessness in a dangerous situation. As Joanna Bourke has argued, emotions like fear "are about power relations. Emotions lead to a negotiation of the boundaries between Self and Other or One Community and Another."[20] A difference in perspective about species recovery leads to different emotions.

An Ice Age Animal Returns to Europe

The muskox (*Ovibos moschatus*), which lives in arctic and sub-arctic tundra landscapes, is a ruminant in the same family as sheep and goats and is most closely related to yaks (figure 3.1). It has a thick coat of wool, called qiviut, which is particularly

Figure 3.1
Muskox at the Tännäs Muskox Farm, Sweden, 2015. *Source:* Author.

warm and water-resistant for the winter and is shed in the sum-
mertime. In the winter months, muskoxen conserve energy
and dig for foliage under shallow snow, but in the summer they
roam in search of grasses, lichens, mosses, and other plants. The
only common natural predator of the muskox, other than man,
is the wolf, so muskoxen use group defense tactics: encircling
calves to protect them and charging enemies with their thick,
imposing horns. Their bodies and habits were designed for the
cold weather conditions of the last Ice Age. Fossil evidence
shows that during the Pleistocene period, which ended about
eleven thousand years ago, their range was circumpolar, but by
the nineteenth century, muskox populations were restricted to
arctic Canada and Greenland.[21]

The first written description of the muskox was published in
1744. Pierre François Xavier de Charlevoix, a French Jesuit and

historian, published a description of "New France" (i.e., North America) in which he described an animal encountered in the Hudson Bay area with long, beautiful hair and a musky smell in rutting season—and he gave it the name *boeuf musqué*.[22] *Musk* was originally the name of the odor from the gland of a male musk deer native to Asia, which was used in perfumes, but animals with similar types of odors were given musk names, like the muskrat and musk shrew. Charlevoix's original descriptor for the animal stuck. It became muskox in English, *moskusokse* in Norwegian, *myskoxe* in Swedish, *Moschusochse* in German, and stayed boeuf musqué in French. The first scientific name of the animal was *Bos moschatus*—an exact translation of boeuf musqué—given by George Zimmerman in 1780. In 1816, Blainville placed the muskox taxonomically into its own genus, *Ovibus*, which represents it as an intermediate between sheep (*Ovis*) and oxen (*Bos*), but retained the musk part as *moschatus*.[23] From a scientific naming standpoint, credit to Blainville for this decision has been overlooked in the scientific literature: the most common scientific name for the species is *Ovibos moschatus Zimmerman*.[24]

The muskox name posed problems at times for those seeking to domesticate and acclimate the animal because of its associations with oxen and musky smells. The polar explorer Vilhjalmur Stefansson noted that muskox meat was quite similar to beef. He said it could be a commercial success "only [if] the name can be changed into something more attractive. ... Although musk was a delicate and expensive perfume as recently as the time of our grandmothers, the fashion has so changed since then that the odor is now known by name only and the impression has begun to spread that it is a stink rather than a perfume. ... Our first problem in domesticating the musk ox is, accordingly, choosing for the beast a new name."[25] Stefansson decided that *ovibos* would make the most sense because it was the scientific name already in circulation for the animal. Later, Alfred Henningsen, a Norwegian involved in trying to domesticate muskoxen, would propose calling the animal a "polar sheep" (*polarsau*).[26] It seems few thought about using the name of the animal used by the indigenous populations of Greenland (*umimmak*) or Alaska (*oomingmak*), which mean "the long-bearded one," although the names were recorded in nineteenth-century travel reports like Sir William Edward Parry's *Journal of a Second Voyage for a Discovery of a North-west Passage* (1824). These names would have drawn attention away from the musky odor and refocused it on the wool coat, which has been the primary motivation for domestication attempts. None of these alternate names have been widely adopted, so we still know the animal today as the muskox.

The muskox had been a European resident at the end of the Pleistocene. A prehistoric muskox neck vertebrate bone was found during the construction of the train line from Oslo to Trondheim through the Dovre mountains of central Norway in 1913 (figure 3.2). This find was interpreted as proof that muskoxen had been present in Norway during the last ice age, along with mammoths, which also had recently been identified.[27] Less than twenty years after the prehistoric bone was found, muskoxen would be back in Norway after an absence of several thousand years.

Adolf Hoel—the founder of the Norges Svalbard- og Ishavsundersøkelser, which would later become the Norwegian Polar Institute (NPI), and a major political figure in Norwegian polar claims—decided to import muskoxen from Greenland to Norway.[28] The reasons behind the relocation of muskoxen to Norway

Figure 3.2
Muskox neck vertebrate bones found in the Dovre mountains on display in the Agder Natural History Museum and Botanical Garden, Kristiansand, Norway, 2016. *Source:* Author.

were complicated, including Norway's polar imperial ambitions, the potential for a new meat supply in a relatively barren land, and the potential for a new wool source.[29] There had been talk of the muskoxen as meat and wool providers in some of the early reports on Hoel's muskox project, but these never became regular uses of the animals, which to this day are protected from hunting.[30] Instead, Hoel would more consistently frame the return of muskoxen as the recovery of a lost part of Norwegian nature, as well as a demonstration of the Norwegian commitment to wildlife.

Hoel hoped that bringing muskox to Norway would smooth over political troubles with Denmark through environmental conservation. Norway and Denmark were locked in a debate over the control of Greenland. In 1814, the Treaty of Kiel required Denmark to cede Norway to Sweden, but Denmark retained all other territories, including Greenland. When Norway became an independent nation (free from Swedish rule) in 1905, Norwegian officials started discussing whether Greenland should be part of Norway. When Denmark claimed sole sovereignty over all of Greenland and its waters in 1921, Norway objected. Norway believed they had a claim to resources in East Greenland, where Norwegian seal hunters had been operating since the nineteenth century.

Norwegian activities on the disputed East Greenland territory came under scrutiny, particularly in their relationship to muskoxen. Norwegian hunters who were hunting arctic foxes and seals for fur in East Greenland often killed muskoxen to feed the meat to their sled dogs. In addition, there had been a high demand for muskox calves to send to zoos around the world— and catching calves required killing the adults. The muskox population showed signs of decline. In December 1926, Landsforeningen for Naturfredning i Norge (Association for Nature Protection in Norway) wrote an open letter to the Norwegian Foreign Affairs Department, criticizing the muskox-killing practices in East Greenland.[31] They pointed to Canada's move to establish the large Thelon Wildlife Sanctuary to protect muskox. The group recommended that a clause about preserving rare or useful species be added to the pending agreement between Denmark and Norway on Greenland. Then, in 1929, the report of the eighteenth Scandinavian Naturalist Congress, held in Copenhagen, was released. It declared: "Now the time has come, which we fear, that muskoxen will be extinct in East Greenland north of Scoresby Sound, and consequently the governments of Denmark and Norway should immediately take measures to protect the muskox. And why should Denmark and Norway not be in agreement about this issue and follow the stellar example which is set by Canada, where the muskox's existence was also threatened, since both its area of distribution and number of individuals was greatly reduced?"[32] The scientific community thus was inserting itself into the international political debate over natural resources in Greenland, claiming that muskoxen belonged in Greenland and that conservation measures were necessary to keep it there.

This is the political context within which Hoel, as head of NPI, decided to catch muskoxen in East Greenland and release them on the Norwegian-claimed island archipelago of Svalbard in 1929. The political implications were clearly on his mind. In 1930, Hoel wrote that the muskox translocation "has national meaning for us. We Norwegians are a hunting people, more than anyone else on the whole Earth. We often hear critique that we exterminate the ocean's mammals, whale and seal, and fur-bearing animals like arctic fox and polar bear. This critique hurts us in many ways. This attempt with transplantation of muskoxen can partially answer this critique; we will show with it that we don't only slaughter, but that we too support cross-border idealistic cultural work."[33] The movement of muskox then was about more than just a source of meat or wool; it was a form of hope through environmental conservation. Hoel argued that preserving muskox was part of his motivation for the rewilding, and conservationists were early supporters of the project to bring muskoxen to Norway.[34] Although Hoel continuously reaffirmed that Norway should be committed to preserving muskoxen on Greenland, a total protection of muskoxen there was unacceptable because the animals were needed by hunters. Hoel argued numerous times that the muskox population on

Greenland was not, in fact, threatened by Norwegian hunting.[35] Despite this belief, he still wanted to build up a population in Norway to show that Norwegians were conservation-minded. There was a longing to do the right thing.

In 1929, Hoel organized an expedition to East Greenland on the ship *Veslekari* to capture muskox calves for relocation to Svalbard. The Norwegian media followed the story closely, with numerous articles appearing from July to October 1929 to inform the public about the exciting project. Hoel himself was unable to lead the expedition as first planned, so NPI geologist (and later head of NPI) Anders Orvin went in his stead. The muskoxen caught in Greenland by the *Veslekari* crew plus some additional calves and young animals that had been brought back by another ship that summer were released on Svalbard in September 1929. When Hoel described the project to the *Aftenposten* newspaper, he stressed the future value of the muskoxen on Svalbard as a meat source for the local population.[36] That was a hope for the future—but in the present, Hoel asked the local mining company to keep an eye on the muskoxen and to instruct their employees that it was forbidden to shoot them. A sign was set up in the mining company's shop in Longyearbyen, stating: "Muskoxen are protected."[37] Hoel's hope was that eventually the population would grow so they could become the promised meat source. But the dream of muskox meat on Svalbard never became reality, and the population slowly declined from natural causes; the last Svalbard muskox was spotted in 1985.

After the first muskox shipment was sent to Svalbard, Hoel arranged for a second group of ten young animals to be released into the Dovre mountains of Norway proper on October 9, 1932.[38] The animals had been captured in East Greenland, transported to Norway (figure 3.3), then held on the island of Ålesund for

Figure 3.3
Muskox calves onboard the ship *Polarbjørn* on their way to Norway from
East Greenland, 1931. *Source:* Adolf Hoel/Norwegian Polar Institute.

about a year before their release at the Hjerkinn train station in
the Dovre mountain area.[39] The project was paid for primarily by
the Landbruksdepartementet (literally, "land use department"),
which had jurisdiction over agriculture and natural resources.[40]
A group of locals came out to the station to catch a glimpse of
the strange new inhabitants. The older animals were quick to
dash off into the hills, but the youngest ones were so tame that
they ate hay out of the hands of the onlookers.[41] The young animals stayed relatively close to a nearby farm, and a status report
in November 1932 made a point of this near-domestic state: "It

is amusing to see the animals coming straight down the mountain to the farm around 5 or 6 in the evening; just like other domesticates. In the morning—at the 9–10 hour—they begin to head up higher into the mountain again. It looks like, any case, that this flock is doing well here."[42] These wild animals seemed to be on the edge of domestication yet were going to roam free in the wilderness.

Their domestic qualities were soon forgotten; the muskoxen brought to the Dovre mountains were framed as the recovery of a lost part of Norwegian nature. The idea that muskoxen belonged in Norway because of their historical presence there, as evidenced by the skeletal remains found while digging the Dovre railway line, was a significant factor in the decision to bring muskoxen to the mainland. When Hoel wrote about the reintroduction plans in 1929, he noted that though the meat-producing potential of the animals was a factor in the reintroduction decision, it was even more important that "muskoxen had lived in our land under or just after the end of the Ice Age."[43] For Hoel, this was not a typical acclimatization effort with an animal species from another part of the world, but instead the return of a species that belonged in Norway. Although it had been thousands of years since they last roamed in European mountains, Hoel expressed hope that the animals would again find their place at home.

There was great hope for muskoxen to fill the Norwegian mountains, but reality was fraught with difficulties. On April 27, 1934, five of the ten animals were killed in an avalanche. In 1938, another two calves from Greenland were released to supplement the herd.[44] Then during the German occupation of Norway during World War II, the herd disappeared. The animals were likely picked off in poaching activities; in one documented case from 1944, a German soldier was accused of shooting a muskox bull.[45] Nazi soldiers stationed in the area to patrol for Russians who had escaped from local prisoner-of-war camps decided to shoot a muskox bull that had been seen near Hjerkinn station. When some local residents saw the soldiers headed out into the mountains with their guns, a man ran after the soldiers to tell them that muskoxen were protected animals. The message was conveyed to the soldiers, yet some minutes later shots were fired at the animal. According to a pair of sixteen-year-old boys who were serving as guides for the soldiers, the junior officer shot the muskox after it had begun to charge the group, which had crept within ten to twenty meters of the animal. This was, of course, only one animal; yet we know that more must have been killed—either for sport by the occupation soldiers or for food by the local population. By the end of the war, the muskoxen were gone from Dovre.

Yet after the war, NPI decided once again to import muskox calves from Greenland to the land where they "belonged": eight animals were released in 1947, and more were added every year from 1949 to 1953.[46] This time, there was no talk of muskoxen as meat or wool providers in NPI's reasoning behind this second set of releases: "The relocation of the animals is of great scientific interest. For the district they can be a real attraction for tourists and for the local population. And conservationists over the whole country have gladly received the news that we were able to acquire calves. The Germans shot the whole old population, which had grown in total to 15 animals. Let us now safeguard these new additions to the high mountain's fauna."[47] As this passage demonstrates, the talk of muskoxen for practical consumptive use had disappeared. Now the purpose for the animals was scientific, environmental, and touristic, which makes the muskox story a historical mirror of contemporary rewilding efforts. The muskoxen were entangled with the rebuilding of Norway as a nation after the German occupation. They were framed by NPI and its spokesmen as an integral part of the Norwegian high mountains; they belonged there. The value of muskoxen was in their presence: they were appreciated as an integrated part of nature and were valuable culturally through tourism, and their return was "gladly received." The emotional response to the muskoxen was all positive on the side of those who brought the animals to Norway.

Conflict in a Wilder World

Not everyone, however, was pleased to welcome these new mountain dwellers. On a national level, the authorities in Oslo looked to the muskoxen as a source of hopeful futures for the Dovre mountains, but locals felt fear rather than seeing hope in the animals. Fear, which is not an irrational response, needs to be recognized as one of the major emotional responses to the recovery of large animals.[48]

Two weeks after the Dovre release, a poem was sent in to the newspaper *Aftenposten* and published under the name *Nicolette*.[49] In the poem, Nicolette noted that she had been a regular hiker in the Dovre mountains and one of the imminent dangers had always been cattle bulls allowed to graze freely. Now she had "heard with a shock that the danger steadily increases. Now someone has let loose a flock of wild muskox." With "muskox in front and muskox behind," she was afraid. "Think," she wrote, "about when a tourist comes carefully creeping! And think when all the bulls in the mountains come into rut!" Nicolette decided that she had made her last trek in the Dovre mountains. She concluded that the danger of "a boxing match with that kind of muskox ruffian" would be too great to risk. In her poem, a palpable fear of meeting a muskox affects Nicolette's actions.

The Norwegian Trekking Association (Den Norske Turistforening, DNT) was concerned enough about these fears that it sent a letter of inquiry in December 1932 to the government, asking whether muskoxen would be dangerous for tourists. The official response from Hoel was that the animals were "not dangerous, but people should be warned not to come too close if one is leading a dog."[50] Although much of the muskox story is dominated by a few central actors who arranged to bring the animals from East Greenland to "improve" the countryside, Nicolette's poem and DNT's concern reminds us that local inhabitants and the people who regularly visited the mountains could see the animals as a threat rather than a promise.

The muskox project proponents regularly stressed the harmless nature of muskox in the press. This is evident from the very first article about the Dovre muskoxen published in *Aftenposten*, carrying the secondary headline, "There is no danger for mountain hikers."[51] Whenever farmers would complain about muskoxen entering their fields, representatives of the NPI would proclaim the animal's peaceful nature.[52] Anders Orvin, subdirector of the NPI, made one such statement to *Aftenposten* in 1950: "Muskoxen are normally totally harmless, and if they have been threatening, the blame is surely that a bunch of people have gone too close to them, or that they have been bothered by dogs. The normal ox that grazes in the mountain fields in summer is more dangerous than muskox."[53]

Yet encounters with muskoxen throughout the mid-twentieth century proved that the fear of muskoxen was not unfounded. Only five years after the first calves had been released, Hoel received a letter from Jacob Skylstad, the editor of the Trond-heim newspaper *Nasjonalbladet*, saying that the muskoxen were known to come into fields with cattle or horses—which was quite disturbing to the farm animals, although so far they had caused no real damage.[54] Skylstad noted instances in which muskoxen had attacked people and pets, particularly when people had approached closely to take photographs. One such tourist ended up running for his life from a charging muskox bull.

Dangerous charging muskoxen were not uncommon in the two decades after World War II. These stories were always associated with tourism or curiosity about the animal. In 1954, a family going into the mountains to pick flowers saw a herd of muskoxen (with calves) and decided to approach them to take pictures. They got within thirty meters before one charged them. No one was injured, but the action photo of a muskox running toward the camera showed it must have been a pretty scary experience.[55] In September 1963, a man tried to photograph a lone muskox that had wandered into the village of Soknedal and was thrown up in the air by the muskox's horns.[56] That animal had been around the village several days, during which time it had been roped on one foot and had stones thrown at it, so it must have been very tired of being bothered. Later that same month, another local man tried to take a photograph of a muskox that had made its way into Tolga, higher up in the mountains. He supposedly got within fifteen meters before the animal charged. Although the photographer was knocked down, he was able to recover his camera and take a picture of the muskox standing below two of his comrades, who had climbed up a tree.[57] The newspaper articles presented these stories as moments of fear—a family running for cover, men clinging to a tree for safety, a man being thrown into the air. The persons who reported the stories expressed their fear of the muskoxen, and that fear was not unfounded.

In 1964, a muskox attack proved deadly for the first time in Norway. Ola Stølen, a seventy-four-year-old farmer, noticed a muskox near his property in Stølen, a little hamlet in the Åmotsdalen valley of the Dovre mountains of Norway. He told his son, daughter-in-law, and grandson about the visitor. The group wanted to observe the seldom-seen animal close up, so they crept to within forty meters and stood still, observing the animal for ten minutes. After a while, the muskox began to snort and then charged the group. Ola was knocked down as the three younger adults quickly ran back to the house. When Ola got up, he grabbed a stone to try to scare away the animal, but the muskox charged again and this time gave Ola a life-ending blow to the chest. Ola's son, who was one of the three onlookers, called the sheriff, Oddmund Hoel. The sheriff in turn contacted the regional muskox manager, John Angaard in Dombås, and asked if they could kill the bull. Angaard replied he would come to take care of it, but before he could get there, the sheriff and Ola's son shot the animal dead (figure 3.4).[58]

The local response to the incident showed both fear and anger. The farming community of Engen held a meeting and drafted an ultimatum to the Landbruksdepartementet, which was signed by 122 residents and sent via telegram: "The under-signed, all residents of Engan, Oppdal, want to make the Ministry aware that the tragic event of Wednesday, 22 July when O. Stølen was killed by a muskox on his property, has created deep unease among folks here. We are therefore bold enough to ask the Ministry to make sure all muskox are removed from the valley between Åmotseiven and Driva by Monday, 24 August 1964. After this date, all muskox which are found in the said

Figure 3.4
Picture of the sheriff and Ola's son with the dead muskox. Published in
Aftenposten, July 23, 1964.

area will be shot. We hope and believe you will understand our
reaction."[59] According to interviews in the papers, the deadline
was picked because it was the first day of school. Children had to
walk three to four kilometers through woodland to school, and
muskoxen had been seen in the area. The parents were especially
worried because it would be dark both before and after school
by late fall.[60] Because many of the farms used summer pasture
and young women were often the herders, there was also a fear
of sending them out alone, both for the sake of the women and
the herds.[61] The residents expressed fear of encountering large

animals on dark pathways. For them, wild, uncontrolled nature was not a hopeful sight but a fearful one. In the minds of the locals, muskoxen did not belong in the landscape.

The Landbruksdepartementet responded that a permit was required to shoot a troublesome muskox and that a general sanction wasn't possible.[62] Action could be taken in an emergency situation during an attack, but otherwise a specific permit was required.[63] Senior Landbruksdepartementet official Helge Christensen made the point that there were other animals in Norway like bears that had killed people, but the entire population had not been exterminated. The comparison strengthened the position that muskoxen belonged in Norway, just like bears. Christensen noted in a newspaper interview that muskox had been in Norway in prehistoric times, and after importation had "become an element in Norwegian fauna."[64] In this statement, Christensen stressed the belonging of the muskox in the landscape and downplayed potential feelings of fear.

The local community was unsatisfied with this response, and at a second group meeting in September reiterated their demands for muskox removal, but no official action was taken.[65] Nor does it appear that the local community started shooting all the muskox. There were animals killed on occasion over the next few years, but always with the appropriate permission according to reports. Incidents involving tourists and photographers continued to crop up in the news reports, but no more deaths by muskox have been reported in the Dovre mountains.

Unbounded Animals

Current proponents of rewilding tout the desirability of self-willed, uncontrolled nature. The hope is that an unfettered nature will recover ecosystems and biological processes lost under human management. But loosening the ropes holding back nature also means accepting that nature might go where it wants to go, rather than where we think it should.[66] This is precisely what happened in 1971 when a small herd of muskoxen from Dovre crossed the mountains into Sweden. When these muskoxen crossed a political boundary, decisions were made about whether they belonged in this space. The emotional responses to the muskoxen's self-determined Swedish rewilding ranged from hope to fear.

The news that muskoxen had come to Sweden was announced in the regional newspaper *Östersunds-Posten* on September 1, 1971, with a warning: "Wild muskoxen in Härjedalen. Warning: They can attack!"[67] Seven individuals (five in a herd plus two lone older males) had been spotted by residents near Funäsdalen, just on the Swedish side of the border in the Rogen mountains. Both the death in Åmotsdalen and the threat to school children were specifically named as examples of the muskox danger. Fear was a first response.

Although mountain hikers staying in the area were "nervous" about the visitors, according to news reports, the news of the herd's relocation also generated optimism among some in the tourism sector.[68] On September 3, the Hamrafjället tourist hotel already had a sign outside advertising an expedition to "Muskoxen's land" (*Myskoxarnas land*) the next day.[69] By the eighth, a national newspaper reported that the local hotel had been "invaded by tourists, who with fear-filled delight are awaiting the great moment" of seeing the animals in the wild.[70]

The herd of five—one male, two females, and two calves— had not come far into Swedish territory, so there was much speculation about whether they would stay in Sweden or move back over the border. One report on September 7 admitted that "the area of Sweden that the muskoxen have visited is very ideal from their perspective," and a newspaper cartoonist humorously interpreted the events of "Norwegian muskoxen rushing toward Härjedalen's mountains" with a running muskox saying "Ah! Cloudberries!"[71] The Norwegian authorities offered no concrete actions to collect the animals and bring them back over the border, instead adopting an attitude of patience to see if the animals would simply migrate back. The manager of the Norwegian Directorate for Wildlife and Fisheries likened the muskox to all other wild animals that migrated over the border and insisted that authorities "do not interfere with the animals' natural wandering."[72]

The animals stayed on the Swedish side of the border and began to multiply: in 1972, two calves were born to the two females in the small flock; by 1976, there were eleven animals in the herd.[73] The owner of the tourist hotel Fjällnas, N. G. Lundh, sent a letter in May 1976 to the Swedish EPA (Naturvårdverket), urging the department to "give muskox a chance" with official protection and scientific research.[74] He noted that the animals generated significant interest as photography subjects and that the Rogen mountains could be "muskoxen land" (*myskoxelandet*). He expressed hope in the muskox as a part of Swedish fauna.

Bengt Andersson, the mayor of the local indigenous Sami village, took a contrary position in a letter sent to Jämtland county in August 1976.[75] Andersson described the problem of muskoxen from the Sami perspective, saying that "the Sami village's grazing grounds are suffering significant inconvenience and intrusion." First, tourists were coming into the area to try to see the muskoxen, causing the Sami's reindeer herds to be disturbed. In addition, the police had used a helicopter in April 1976 in an attempt to drug a muskox that had been injured, and the operation disturbed the annual reindeer collection, with the result that two extra weeks were required to herd all the reindeer that had been scared away. In Andersson's comments and letters, he often used the word *olägenhet* to describe the muskox situation— the word has a connotation in this context similar to "nuisance" in English law. In other words, it's something that bothers you enough to keep you from doing the things you think you should be able to do. But he also uses *farliga*, meaning "dangerous." So muskoxen in Tännäs were both inconvenient and dangerous to the reindeer herders. To limit that inconvenience and danger, the Sami community wanted to make sure no more muskoxen would come into the area, and they wanted financial compensation for their losses.[76]

This growing realization that muskoxen were causing conflict prompted the Swedish EPA to organize a symposium in Funäsdalen to bring together key stakeholders on September 2, 1976.[77] Representatives from the local governments, police operations, the natural history museum, tourist interests, and the local Sami village participated.[78] All of the stakeholders agreed that having a few muskoxen in Sweden was not bad, but serious concerns were raised about a large population and the potential conflicts with local land use. On the agricultural side, there had been problems with muskoxen going into pastures, eating hay, and even breaking fences.

Andersson offered the most criticism during the meeting, detailing the reasons for his skepticism. Because Sami livelihood depended on semidomesticated reindeer, which were allowed to roam in the mountains and were then herded at the end of the season, muskoxen were seen as a significant problem. On a practical level, the animals had been observed within the reindeer flocks, which made it difficult to herd the reindeer using dogs because muskoxen tended to charge dogs. There was also damage to mountain cabins and fenced areas that was attributed to muskoxen. In addition, both the tourists and naturalists searching for the herd by plane, snow scooter, and foot disturbed reindeer during the calving season, leading to abandoned newborns. Andersson wanted guarantees that no additional muskox would be allowed in the area and questioned whether the ones there now should be allowed to stay. He also demanded compensation from the national or local authorities for costs incurred because of muskoxen. In his comments, Andersson used words like "worried," "very difficult," and "forced to take the long way" when describing how the Sami population experienced the muskox presence. Although muskoxen might be tolerated to some degree, Andersson did not believe they belonged in the landscape.

The others in the room dismissed Andersson's complaints on three grounds: that muskoxen (1) are a "nature protection" object, (2) can bring tourism money to the area, and (3) aren't any more dangerous than other kinds of wild animals, such as moose. The draft version of the Swedish EPA's proposed muskox policy issued at beginning of 1977 had to deal with both sides. The policy's opening statement recognized the tension between hope and fear in this unintended rewilding: "muskoxen shall be protected as a part of the Swedish fauna but be subjected to careful control."[79] The policy included provisions for capturing and returning muskoxen which had left the flock, strongly suggested that the county-level administration designate a person or group to produce regular reports on the animals' whereabouts, ordered the creation of informational brochures on the animals, and encouraged research on muskoxen. Protection for the animals was considered to be covered under the existing hunting laws, but there was a recommendation to change the law to offer financial compensation for death or injury by muskox. The policy also specifically mentioned that air traffic during the muskox calving should be reviewed.

After some minor textual changes, a draft of the policy was sent out for public comment in August 1978. There were several critical responses. For example, the veterinarians of the Agricultural Board (Lantbruksstyrelsen) and scientists at the Swedish Agricultural University and the National Natural History Museum complained that the preference for drugging and returning animals to the herd was not scientifically sound because these animals had likely been rejected by the herd.[80] Most importantly, the local community was not pleased with the policy's failure to severely limit the number of allowable muskoxen in the country and its lack of financial compensation. The Sami community, in particular, wanted a much stricter policy: the number of allowable muskox should not be above twenty; absolutely no muskox monitoring expeditions between April 20 and June 15, when reindeer calve; a financial commitment to remove muskoxen that move into reindeer flocks; and explicit governance involvement from the local Sami township.[81] The Sami were not alone in demanding a small population; the Jämtland County Board also stated that a limit of around 20 animals was needed.[82] The Jämtland Agricultural Board stressed that compensation for more than just death or injury should be included. They recounted incidents of a mother and child in the summer of 1978 being threatened by a muskox and an attack on a dog that ended the dog's life as examples of events that should be compensated.[83] These particular histories stressed the fear experienced by the local population because of the presence of muskoxen.

Despite these objections, the official government position issued in May 1979 did not change. The final policy still listed "two complete herds" as the desirable limit, although the language was softened to a preference. No special compensation packages were included, nor was there any mention of possible damage to farmers and herders, although the memorandum did recommend that human death and injury by muskoxen be added to the compensation law. In addition, the memo recommended adding the muskox as a legally protected species.

This story shows that even when local input on environmental policy is requested, it's easily shoved aside. The Sami views of what should be done with the muskox did not match the "desired" outcome of the conservation or natural science communities, so it wasn't incorporated into the muskox policy. The muskox had been positioned in the policy as a species that belonged in Sweden, even though it had been absent for thousands of years before a herd of hungry animals roamed a little too far over the mountains.

The muskox herd numbers climbed through the 1970s and 1980s, up to a high point of thirty-five individuals. In fact, the muskox become a figurehead of sorts for the Swedish border mountains. When a Swedish national postage stamp series titled *Fjällvärld* (Mountain World) was issued in March 1984, the images chosen were the angelica flowering plant (also known as wild celery), the lemming, and the muskox. This human inclusion of muskox in the Swedish fauna came only thirteen years after the herd had immigrated over the border. The text printed (in both Swedish and English) on the first-day-issue card for the stamp series reveals the rapid integration of muskox: "In 1971 the musk-ox (Ovibos moschatus) came back to the Swedish fauna. The occasion can be seen as a return to the fold, and today there are some 30 animals in the province Härjedalen."[84]

In this text, the muskoxen coming to Sweden was "a return to the fold," or a return home. The idea was that muskoxen were native Nordic animals that had at last come back to Sweden.

Had muskoxen won a stamp of approval? Perhaps not. When the muskoxen suddenly started disappearing in the late 1980s, poaching was specifically blamed because "it's no secret that many here in the area want the muskoxen to be gone."[85] Clearly not everyone had bought into the hopeful narrative of the muskox. The actual cause of the decline is unknown—it could have been a livestock disease similar to the lung inflammation disease that struck the Norwegian herd in 2013 and wiped out over a quarter of the population in one year—but it's apparent that the muskox was understood by some locals as a creature that did not belong.

As the small Swedish herd entered a dramatic downward population trend, some direct recovery actions took place. First, supporters of the muskox released a younger male to take over a herd that had an aging male and no young competitors to take his place. Then, they attempted to get the animal listed as endangered, which would have seemed to be the natural outcome of the 1979 policy that proposed protection for the animal. In 1990, a parliamentary motion was filed by four Green Party members to request an investigation into potential ways to help the muskoxen in Härjedalen, arguing that because paleoarcheological finds of muskox had been made in Sweden, "the species belongs truly to Sweden's original inhabitants. ... We have no right to abandon the muskox."[86] The official parliamentary response to the motion shows that everyone was not in agreement about the muskox's place in Sweden: "Musk oxen have been extinct for so long in Scandinavia that they can no longer be considered as part of our natural fauna [vår naturliga fauna]. However, they are without doubt a particular interest in the tourism industry. The existing herd in Härjedalen is so small that there is a risk of degeneration and extinction. A strengthening of the herd would, according to the experts, be costly and cumbersome and take a disproportionate share of the available conservation funds."[87] A second parliamentary motion made in 2000 and a draft of a Threatened Species Action Plan for the muskox were both denied.[88] The new official position had declared that the muskox, although in Sweden, did not actually belong there and was not eligible for endangered species protection.

Despite these setbacks, environmental groups have continued to support the muskoxen population. In 2013, the nonprofit muskox breeding and visitor center Myskoxcentrum in Tännäs released a young breeding female to join the remaining six animals to strengthen the breeding potential of the herd.[89] The center is still pursuing the dream of a vibrant muskox herd in the Härjedal mountains. Hope is the dominant emotional response from members of environmentalist groups and tourism organizations when discussing the muskox.

Controlling the Wild

The group of muskoxen that crossed over the border to Sweden in 1971 were not the only free roamers. Newspaper accounts in Norway regularly tell of animals showing up uninvited and unwanted in the center of towns, on roads, in camping spots, on train tracks, and in tunnels.[90] Muskoxen had to be forcibly removed or killed if they strayed into human-inhabited areas.[91] In 1983, the Norwegian population was thirty-six-muskoxen strong—but that would grow significantly to eighty to ninety animals in 1993, so the potential for transgression into unwanted spaces was growing.[92] With those transgressions came fear and questions about where muskoxen belonged.

In 1988, the Norwegian Ministry of Environment suggested that a management plan for the Dovre muskoxen was needed and delegated the task to Sør-Trøndelag County officials. Their responsibility was to develop a plan to define limits on "spreading the species, how large the population should be, and how the population should eventually be regulated."[93] In 1994 before the calving season, there were fifty to sixty-five animals, staying generally in one area of the Dovre mountains. Within this framework, the management plan defined a "core area" and a "stray area" within which muskoxen were permitted to roam, included a decision to limit the population to fifty to seventy animals, and developed a procedure for handling human-animal conflict. The core and stray areas do not align with topography; populated areas and political boundaries were more influential in the mapping exercise than habitat.[94] Animals outside of the allowable areas would require removal or culling to avoid conflict with humans.

The document created an argument for the muskox's belonging. Muskoxen were positioned as "a part of the Norwegian fauna," even though the ancestors of the present animals were brought to the area only in the 1950s and had been classified by the Directorate for Nature Protection as a non-native introduced species.[95] The report argued that "although the muskox has been in the Dovre mountains for a relatively short time, it has been strongly tied to the mountain area in the popular imagination. In many ways, muskox has become a symbolic part of the Dovre Mountains National Park."[96] The management measures were thus premised on the position that muskoxen should be in the Dovre mountains "as long as possible."[97]

The plan has been updated each decade. In the 2006 version, the core area was expanded from 169 km^2 to 340 km^2 to match the area that the muskox herds had actually been using since the 1996 plan.[98] By then, there were 170 animals, and the report retracted the "limit" on the number of muskoxen (which obviously had not actually been implemented over the prior decade). In the next iteration of the report, which was finalized in 2017, the core area remained unchanged, but the report again implemented a limit—this time specifying two hundred overwintering animals.[99] A new suggestion was made to allow for private hunting of muskoxen to control the population. The muskox had established itself as an integrated, yet thoroughly managed part of the Norwegian mountainscape. It belonged, but only in certain places. Fear would be countered by control.

Rewilding the World

An updated vision of "Dovre Mountains' newest ornament," as the 1940 newspaper article that started off this chapter had labeled muskoxen, was unveiled in 2015 by the Norwegian artist Sverre Malling, who spent six months on a large format (180 × 263 cm) charcoal drawing titled *Norwegian Muskox*. The huge shaggy muskox stands on rocky ground, dominating the artic mountain landscape that shrinks behind it. Its long hair flows down the body; wool tuffs cover the top in patches as if the animal has been recently shedding. Its gaze to the left is away from the viewer, yet the stare penetrates the scene. The muskox stands ready as guard over the landscape. Tucked away in the lower left is a traditional rural cabin with its wooden walls, living grass roof, slanted wooden fence, and farm tools against the sidewall. The cabin seems insignificant, overshadowed by the looming large beast.

The image encapsulates the contradictions of the rewilding of Scandinavia with muskoxen. Great hope was placed in these animals—first by Adolf Hoel and the Norwegian Polar Institute, and later by tourism businesses and environmentalists. They hoped to bring back something wild to the peninsula's landscape. Muskoxen, which would be free to roam and multiply in the relatively sparse mountainous landscape, were understood as a unique historical component that had been lost. They had been found, and that brought hope. Yet when the lost nature is restored, other things may be lost. Historically, great fear was attached to these animals by mountain trekkers, rural farmers, and indigenous herders. Allowing muskoxen to roam free and multiply came with costs—real costs of damaged property, lost time, and even lost lives. Those fears were not unfounded.

The many ongoing rewilding consultations happening across Europe at present face the same kind of contrast between hope and fear. Distant environmentalists, often upper-class white urban dwellers, are vocal about their desires to rewild the countryside and imbue those activities with great hope for a recovery of nature. Yet local residents, the ones living and working in the countryside, are the ones that have to adapt to the muskoxen or wolves or bears in their daily lives. Emotional attachments on both sides motivate their positions, just as they have in the Scandinavian muskox story.

In October 1975, Randi Bakke wrote a letter to *Aftenposten* after a second fatal attack by a muskox, which left a moose hunter dead in northern Norway. The letter, which ran under the headline "Are we benefited by muskox in Norway?" asked some key questions about rewilding with muskox: "What will we really do with these imported animals? Do they have any justification at all in Norwegian nature? They do absolutely nothing valuable ecologically, nor are they meant to be used as edible game. They are here only for enjoyment, for whom?

Sure, for some enthusiastic people who think it is fun (exciting) to know that we have some flocks of this animal running around in our mountains. Moreover, it's newspaper material for the masses! ... Are we benefited by it? Does it have some sensible purpose? Who will or can answer that?"[100] Randi's questions get to the heart of the matter: Why bring in muskox and set them free in the Norwegian mountains? She wanted a concrete reason grounded in economics or utility or even ecological benefit, but in so doing, she overlooked the power of hope as an emotion. The rewilding enthusiasts hoped (and still hope) to make a wilder world. This is not a hope based on a present and immediate gain. Hope functions as bridge "between the beliefs and actions of today and possibilities for tomorrow," inspiring environmental actions that may lead to recovery in the future.[101]

4 Resurrecting: Grieving the Passenger Pigeon into Existence Again

In February 2012, a group of scientists, environmental thinkers, and science writers gathered at Harvard Medical School. They were not there to talk about medicine to heal a wound or illness, but rather about bringing the dead back to life. The daylong meeting explored the technical feasibility and potential pitfalls of using genetic engineering to resurrect the extinct passenger pigeon, *Ectopistes migratorius*.[1]

The Long Now Foundation, a nonprofit cultural institution, sponsored the meeting, which was called *Forward to the Past*. Stewart Brand, known for his *Whole Earth Catalog* aimed at supporting American counterculture and the nascent environmental movement in the late 1960s and 1970s, established the foundation in 1996. Brand had been an early advocate of the power of thinking big: he famously asked, "Why haven't we seen a photograph of the whole earth yet?" on buttons in 1966 and worked toward making the image available.[2] The foundation is attempting to instill long-term thinking in modern culture through projects like the Clock of the Long Now (a clock designed to run ten thousand years without intervention) and the Long Bet Project (which takes bets on predictions for conditions no less than two years into the future).[3] Before the group gathered to talk about the passenger pigeon, the foundation had already shown an interest in the potential extinction of human language: the Rosetta Project produced an engraved metal disk holding fourteen thousand pages of information and has collected texts and recordings in over 2,500 languages to preserve them for the future. The passenger pigeon meeting would kick off the Long Now Foundation's interest in nonhuman extinction, which after the meeting was formalized in the foundation project called Revive & Restore.[4]

The meeting centered on rapidly advancing genetic technologies that allow the recovery of ancient DNA sequences, as well as detailed genome editing. The combination of these two advancements would mean that DNA of long-dead species could be sequenced and then remade by modifying the DNA of existing species. George Church, a genomic engineer at Harvard Medical School who served as the on-site host, stressed the high degree of precision in genome-editing techniques and proposed that the band-tail pigeon *Patagioenas fasciata* could be used as the starting genome for making a new passenger pigeon. The cultural side was less represented than the technical, but Noel Snyder, who had worked on the California condor recovery project, warned the participants that conditions that had led to the condor's endangerment had not been corrected and that "politics can often overwhelm everything else that you are trying to do."[5]

The people around the table apparently considered the passenger pigeon as a self-evident candidate for de-extinction. Although other candidates for revival were mentioned as potential alternates, including other recently extinct North American birds (Carolina parakeet, heath hen, ivory-billed woodpecker, Labrador duck, and great auk) and the wooly mammoth, no one questioned why the passenger pigeon was being discussed in the first place. The history of the passenger pigeon's extinction—a dramatic decline from billions of birds to none in half a century—was mentioned briefly by Joel Greenberg, who was then writing a significant monograph on the pigeon, *A Feathered River Across the Sky: The Passenger Pigeon's Flight to Extinction*, published in 2014.[6] The drama of the pigeon's extinction seemed to be enough justification to make de-extinction an appropriate response. The pigeon was lost, and perhaps it could be found.

In this chapter, I will trace why the passenger pigeon in particular was identified as the bird to be resurrected first. In the stories about the pigeon, the bird is woven into the fabric of the North American landscape. In narratives of abundance, the writers claim that the birds *belong* there, primarily for humans to harvest. As the realization of the pigeon's extermination took hold, the narrative shifted to a profound sense of *longing*. The pigeon's demise seems to leave a hole in nature that many writers comment on. As this chapter will show, by the centenary anniversary of the last pigeon's death, the grief over the pigeon had reached the stage of accepting its fate, yet this grief was unsettled by the beginnings of attempts to resurrect the species.

Ecocriticism scholar Ursula Heise has argued that a common rhetoric of decline accompanies species extinction narratives in literature of all types, from science fiction stories to scientific information databases.[7] She has advocated an environmentalist rhetoric beyond the decline-of-nature narrative, complicating even further the form and function of extinction narrative. Here I build upon Heise's insights to analyze the rhetorical strategies used to narrate the passenger pigeon's life and death in a broad range of literary texts, newspaper articles, artistic works, and museum exhibits and to create a holistic picture of the communication of this environmental event. But instead of looking for declensionist or progressivist narratives, I examine the passenger pigeon's history through the emotional framework of grief.

Grief is a complex emotion, appearing in many guises. I will employ psychologist Elisabeth Kübler-Ross's model of the emotional stages of experiencing death, which identified five stages of grief: denial, anger, bargaining, depression, and acceptance.[8] In her coauthored follow-up book, *On Grief and Grieving*, she noted that not all those who grieve will experience all five stages and that the stages may not be experienced linearly.[9] I want to acknowledge that there has been much discussion within the psychology community about how accurate the stages-of-grief theory is.[10] Although there are certainly many various ways of working through grief, as rightly pointed out by critics, I believe there is still value in the concept that there are different, distinct types of grieving, even if these may overlap and do not follow a given order.

Grief is now the preferred scholarly term rather than *mourning*, which Freud used, because grief focuses on an internal emotion as opposed to the outward social behaviors seen in mourning (such as a widow wearing a mourning dress).[11] Grief is also distinguished from *melancholy*, which is a pathological condition featuring profound dejection, cessation of interest in the outside world, and self-reviling.[12] Melancholy is a state of helplessness because of past loss, whereas grief is forward-looking and changes over time. Although grief is often associated with the death of family members or close friends, grief may extend to other kinds of losses, including loss of the environment.[13] This chapter employs the five stages model because of its explanatory power for thinking about the primary emotions revealed in narratives told about the loss of a species.

Denial: A Land Filled with Pigeons

The sheer number of passenger pigeons (*Ectopistes migratoria*) made deep impressions on early Americans. The passenger pigeon was a migratory bird moving across vast swaths of North America, with wintering grounds as far south as what are now Florida and Texas and summer breeding grounds in the Midwestern states up to Ontario. It was primarily a woodland bird, with regular roosting in oak, beech, and chestnut stands. Mark Catesby's early eighteenth-century work on the birds of the southern states described the passenger pigeon coming to the region in "incredible Numbers; insomuch that in some places where they roost (which they do on one another's Backs) they often break down the limbs of Oaks with their weight, and leave their Dung some Inches thick under the Trees they roost on."[14] When John James Audubon, who published an exquisite drawing of a passenger pigeon pair in his life-size *Birds of America* series (figure 4.1), described the passenger pigeon in 1831, their numbers amazed him: "The multitudes of Wild Pigeons in our woods are astonishing."[15] Describing one migration event he witnessed in 1813 in Ohio, Audubon wrote that "the air was literally filled with Pigeons; the light of noon-day was obscured as by an eclipse, the dung fell in spots, not unlike melting flakes of snow; and the continued buzz of wings had a tendency to lull my senses to repose. ... The Pigeons were still passing in undiminished numbers, and continued to do so for three days in succession."[16] Their numbers were "immense beyond conception."[17] The grandeur of the pigeon flocks evoked feelings of wonder and the sublime.[18]

The pigeons in these vast and seemingly unending flocks were remarkable, but not out of place. The new American continent, which was still early in its colonization in the first half

Figure 4.1
Passenger pigeon. Illustrated in John James Audubon, *The Birds of America from Original Drawings* (1827), plate 62.

of the 1800s, was understood as a boundless source of natural resources. The Midwest, where the pigeons were most numerous, featured fertile land that had been parceled out to farmers along the grid. Commercial agriculture centered on grain turned the Midwest into America's breadbasket.[19] From the mid-1800s, new train lines increased the connections between the resource-abundant Midwest and West with the financial centers in the East, making it faster and easier to move commodities.[20]

The abundance of the passenger pigeon led to appropriation of the bird as a natural resource.[21] Even in 1731, Catesby recorded that from New England south to Philadelphia, a great number of pigeons were shot or knocked off their roosting places during their migrations.[22] For Audubon, it made sense that when the flocks came through the area, "the banks of the Ohio were crowded with men and boys" who shot down sacks full of the birds so that "for a week or more, the population fed on no other flesh than that of Pigeons."[23] Puritan preacher Cotton Mather remarked that passenger pigeons were "frequently sold for Two Pence or Three Pence a Dozen: tho' two or three of them, Roast or boil'd or broil'd, may make a meal for a Temperate Man. Yea: they are sometimes kill'd in such plenty, that the countrypeople feed their Hogs with them."[24] When "old-timers" recorded their recollections of wild pigeons in the late 1800s, narratives of plenty dominated. James B. Purdue of Plymouth, Michigan, reminisced in 1894 about flocks that would "darken the air" and "swarm down" so he and his father could catch "hundreds of them in a single morning." The pigeon breasts were pickled in weak brine and hung to dry, creating "a dainty morsel and would last as long as dried Beef, and was far its superior in taste."[25]

Despite its potential value as food, the great number of passenger pigeons posed a threat to the American settlers who began to farm on the frontier. The birds had a diverse diet, including nuts, berries, and insects, but farmed grains like buckwheat were boons for the flocks. James Fenimore Cooper included a passenger pigeon hunting scene in his historical novel *The Pioneers* (1823), which offers both details about the hunt and visions of

its potential for destruction.[26] Set in New York in the Otsego Lake
district, the action of the novel one April revolves around the
arrival of the passenger pigeons. As "a flock that the eye cannot
see the end of" migrated up from the south, the whole community jumped into action to defend their agricultural fields:
"Every species of firearm, from the French ducking gun, with a
barrel near six feet in length, to the common horseman's pistol,
was to be seen in the hands of the men and boys; while bows
and arrows, some made of the simple stick of walnut sapling and
others in a rude imitation of the ancient cross-bows, were carried by many of the latter." The men went up to the mountain
roosting area and began shooting the birds en masse, yet "none
pretended to collect the game, which lay scattered over the
fields in such profusion as to cover the very ground with fluttering victims." When a small canon was taken out to join in the
action, the wilderness-loving character Leather-Stocking indig-
nantly objected to the events: "Here have I known the pigeon
to fly for forty long years, and, till you made your clearings,
there was nobody to skeart or to hurt them, I loved to see them
come into the woods, for they were company to a body, hurting
nothing. ... Well, the Lord won't see the waste of his creatures
for nothing, and right will be done to the pigeons, as well as others, by and by." After his admonition, Leather-Stocking showed
how he could skillfully take down a single bird with one shot.
The settlers were duly impressed with Leather-Stocking, and
Judge Marmaduke responded, "I begin to think it time to put
an end to this work of destruction." Yet the slaughter continued after Leather-Stocking left for home, and the pigeons were
eventually driven away from the area. In Cooper's novel, the
pigeon hunt is characterized as "sport" and "carnage" that kills
the "harmless" and "innocent." The entire scene was portrayed

as a battleground in which the humans were eventually victorious over the birds, but not without moral reproach.

The pigeons were understood as existing in unending abundance on the land. Audubon noted that they "are killed in
immense numbers, although no apparent diminution ensues."[27]
Pigeons were shipped by the barrel to urban centers like New York
on the East Coast for consumption by the urban poor. Although
the kill rates were high, commentators in the mid-1800s saw
no lessening of the pigeon numbers. The text accompanying
an illustration of shooting passenger pigeons in Louisiana published in 1875 likewise described "the immense flocks of wild
pigeons ... frequently breaking and twisting the limbs of the
forest trees as if a hurricane had passed through the woods."[28]
Audubon stressed that only deforestation, not hunting, would
bring about any reduction in the number of passenger pigeons:
"Persons unacquainted with these birds might naturally con-
clude that such dreadful havoc would soon put an end to the
species. But I have satisfied myself, by long observation, that
nothing but the gradual diminution of our forests can accom-
plish their decrease, as they not unfrequently quadruple their
numbers yearly, and always at least double it."[29]

Audubon believed that passenger pigeons might even become
abundant outside of their home territory. In 1830, he bought 350
live birds in New England and took them to England, where he
distributed them among nobles and zoologically inclined gen-
tlemen.[30] Some of the birds were let loose—a common practice
in the 1800s during the heyday of acclimatization movements
that attempted to establish species from the colonies in the Old
World or vice versa.[31] Some of the passenger pigeons apparently
survived, at least to the extent that Gregory Smart listed them as
rare birds found in the wild on the British Isles in 1886.[32] These

free-range birds evidently never bred with any success and died out. Rapidly and unexpectedly, the passenger pigeon back home would die out as well.

By 1894, the local bird expert Oscar Byrd Warren of Palmer, Michigan, was worried about the pigeon because he hadn't seen any in a few years. Over the next four years, he solicited reports about passenger pigeon sightings far and wide.[33] Warren, who worked as a mining engineer, was an avid birdwatcher and collector, amassing a significant collection of specimens and several notebooks full of entries for bird sightings. The people who wrote back to Palmer remarked that the birds had not been seen in a few years, much to their dismay. They offered no conclusions about why the wild pigeons were no longer common.

In their letters, informants always wondered where the pigeons had moved to; the assumption was that they must be somewhere else. For example, H. T. Blodgett of Ludington, Michigan, wrote in 1904: "Much to my regret I have seen none of the beautiful birds for about six years. The savage warfare upon them, from nesting place to nesting place by pot-hunters and villainous fellows who barreled them for market, with nets and every brutal means for wholesale destruction, has driven them, I know not whither. If there are considerable flocks of them anywhere, I should be glad to know it."[34] There was an underlying assumption that though the bird's numbers had greatly decreased, they were not necessarily on the verge of extinction.[35] There was dismay that the bird, which was understood to belong in places where it had been seen before, was missing.

Sightings of passenger pigeons continued to dribble in. In July 1897, the ornithology journal the *Auk* published a sighting of a flock of about fifty in southern Missouri in 1896, and then in 1897 approximately three hundred birds divided into seven small flocks were observed in Wisconsin.[36] A 1898 newspaper article in the *Chicago Tribune* claimed that the passenger pigeons had relocated to Mexico, where they had been driven because of intensive hunting around Chicago.[37] Although one writer admitted that in the Midwest states "large flocks of Passenger Pigeons are a thing of the past," small flocks and individual sightings were still recorded. Even in 1906, John Burroughs, the editor of the journal *Forest and Stream*, claimed that a large pigeon flock was seen in New York State: "I have no doubt, therefore, that the wild pigeon is still with us, and that if protected we may yet see them in something like their numbers of thirty years ago."[38] Extensive correspondence about the sightings in 1905 and 1906 appears to support the claim that the 1906 birds really were passenger pigeons, but they too would soon be gone.

Anger: What Has Become of the Wild Pigeons?

In 1907, W. B. Mershon published *The Passenger Pigeon*, a collection of recollections and publications about the bird. He stated that his intent was "to throw light on the oft-repeated query, 'What has become of the wild pigeons?'"[39] This was a question that puzzled many from the last decade of the nineteenth century. Passenger pigeons, which were itinerant birds that followed no set migratory route or schedule, had tended to show up in a given locality sporadically every few years. Even if the birds had not been seen in a while, the assumption had been that they were simply somewhere else. But by the time Mershon wrote, observers had compared notes and found that none of them had seen the pigeons. The emotional reaction to the recognition of the pigeon's loss was anger, one of the significant stages in the grieving process.

Mershon's introduction to his compiled text claimed that the passenger pigeon was extinct at the time of the writing (in 1907), and the blame was to be placed squarely on Americans, who were "wasteful" and always in "pursuit of the almighty dollar."[40] The bird's natural behavior, including laying only one egg each season and thriving only in large flocks, could not cope with the "the numbers slaughtered by the professional netters."[41] Although theories of mass die-offs caused by natural disasters like hurricanes and unexpected snowstorms appeared off and on,[42] most commentators after 1900 agreed that overhunting in combination with forest clearance was the culprit for the pigeon disappearance. Mershon revealed his anger in the face of the extinction, decrying that nature was "wantonly destroyed with no thought for the future."[43]

Capitalistic exploitation of the pigeon was frequently derided. Chief Simon Pokagon of the Pottawattomie compared the indigenous method of targeting squabs rather than adult birds with methods of the "white men" who started netting pigeons for market around 1840 and "banded themselves together, so as to keep in telegraphic communication" to learn the movements of the flocks. According to Pokagon, "hunters from all parts of the county gathered at these brooding places and slaughtered [the pigeons] without mercy."[44] In a paper read at the Manitoba Historical and Scientific Society in 1905, the naturalist George E. Atkinson characterized the extinction of North American species, including the passenger pigeon, as a "paragon" in the "career of human selfishness."[45] In 1903, Sullivan Cook recorded his remembrances of passenger pigeon from a young age in Ohio. After detailing capture methods, he admitted being "ashamed of the slaughter." "Young men who are now hunting for something to shoot and wondering what has become of our game," he wrote, "must hear with anger and regret such as reports as this one from western Michigan in the days gone by: 'In three years' time there were caught and shipped to New York and other eastern cities 990,000 dozen pigeons.' ... And when you are asked what has become of the wild pigeons, figure up the shipping bills, and they will show what has become of this, the grandest game bird that ever cleft the air of any continent."[46]

Pokagon, Atkinson, and Cook were all angry—furious at the unneeded destruction of the pigeons—but also ashamed or filled with regret that no one had stopped it. Just as corporate guilt had operated in the case of the locally extinct beavers in Sweden, these men regretted the actions of other humans that led to the pigeon's end. Grief is not exclusive of guilt; in fact, feeling guilty about causing the loss of someone/something can be a main motivation behind intense grief. This is a reminder that emotional reactions to extinction and recovery are often multivalent: one can be both angry and depressed at the same time. Grief as an emotion operates that way, with its potentially overlapping components of denial, anger, bargaining, depression, and acceptance.

In these texts, the passenger pigeon is depicted as a creature that was minding its own business at home when invaders from the outside came in. The pigeons in their brooding places were natural and belonged there; the hunters with their telegraph lines and rifles were unnatural and undesirable. The sentiment that men had forced the pigeon out of the land to which it belonged appears also in the claim that the flocks had simply moved to Mexico.[47] The passenger pigeons had supposedly relocated to Mexico because they had been "banished by man's cruelty" from the north. The illustration accompanying the article shows the expected hunters taking aim, but it also shows a canon being fired at the flocks—something that doesn't appear to have been a regular technique according to common documentary descriptions, but does mimic Fenimore Cooper's *The Pioneers*. The choice of image was a clear signal of the wanton destruction. Before this time, illustrations of hunting had been embedded in narratives of plenty, so this was a shift in the representation of those hunting practices.

Anger about the wanton slaughter of the pigeon reached a pinnacle with *Our Vanishing Wild Life*, published by William Hornaday in 1913. Hornaday—who was the chief taxidermist at the United States National Museum (later renamed the Smithsonian), then director of the Bronx Zoological Park for thirty years, and a key early wildlife conservation figure—had been heavily involved in setting up the American Bison Society to save the iconic American animal after it had become extinct in the wild.[48] The pigeon and buffalo were often linked in discourse about extinction, with both as allegories of the dangers of capitalism and the pigeon's extinction serving as a potential outcome for the bison.[49] Hornaday's book, which carried the subtitle *Its Extermination and Preservation*, was intended as "a loud alarm, to present the facts in very strong language, backed up by irrefutable statistics and by photographs which tell no lies, to establish the law and enforce it if needs be with a bludgeon," according to New York Zoological Society President Henry Fairfield Osborn, who wrote the book's foreword.[50] On the first page of Hornaday's preface, he stressed that he was "appalled" at North America's "wild-life slaughter" and "a restless, resistless desire to 'kill, kill!'"[51] The passion and anger of Hornaday's text comes through via the excessive number of exclamation points he uses throughout the book.

The frontispiece facing the title page is labeled "The Folly of 1857 and the Lesson of 1912."[52] On the left is a quote from an

1857 report for the senate of the state of Ohio: "The passenger pigeon needs no protection ... it is here to-day and elsewhere to-morrow, and no ordinary destruction can lessen them." On the right is a picture of a bird identified as "the last living passenger pigeon now in the Cincinnati Zoological Gardens. Twenty years old in 1912." Above the box is a quote from Patrick Henry: "I know no way of judging of the Future but by the Past." Horna-day labeled this entry into his book as a folly and a lesson. The use of *folly* indicates a harsh and moral judgment. He used this judgment to set the tone of his whole book, which chastised the American people writ large for their ruthless extermination of wildlife, including game birds, song birds, big game, and more. In the main text, Hornaday wrote a short summary of the "long and shameful story" of the passenger pigeon's decline and extinction, mincing no words as he outlined how "millions were destroyed so quickly, and so thoroughly *en masse*."[53] He declared that with just one individual known alive, "the passenger pigeon is a dead species."

As the disappearance of the great flocks of passenger pigeons was perceived by Americans at the end of the nineteenth century, remarks spanned from bewilderment to anger. Commentators such as Mershon and Hornaday insisted they knew hunting was to blame for the extinction and expressed anger, even fury, at the failure to do something to stop it. All could agree that the wild pigeon was lost from the place that it belonged.

Bargaining: Attempts to Resuscitate a Dying Breed

In the last decade of the 1800s, there were some attempts to help the passenger pigeon numbers recover. To start, a few groups proposed hunting moratoriums or permanent laws banning hunting of particular types. In Pennsylvania, shooting firearms at a roosting pigeon was illegal from 1873, and subsequent legislation strengthened limitations on hunting nesting birds, although it still allowed trapping elsewhere.[54] In Michigan, ornithologist Ruthven Deane of Chicago proposed a bill in the State Legislature of Michigan in 1896 to ban wild pigeon hunting for ten years to allow their numbers to recover, and the bill became law in 1898. These early legislative attempts at controlling passenger pigeon hunting led the way for the eventual adoption in 1900 of the US Lacey Act (16 U.S.C. § 3371–3378), which was the first significant legislation aimed at the preservation of game and wild birds. But all this was too late to reverse the damage done to the passenger pigeon numbers.

In a more intensive intervention, some maintained that wild pigeons could recover through breeding in captivity. This was the tactic taken successfully for the American bison (known as *buffalo*) and European bison (also called *visent*). The American variety had been reduced in number in the United States to less than one hundred free ranging plus two hundred in Yellowstone National Park by 1889.[55] Private herds, as well as those owned by the federal government, became the breeding stock that were released onto public lands in the first decade of the 1900s.[56] The European bison suffered an even more dramatic decline, with no existing wild population in 1927. Zoological garden populations that had been gathered at the beginning of the 1900s were actively bred through the International Society for Protection of European Bison in the 1920s, and twelve animals became the ancestors of all wild visent today.[57]

Passenger pigeons were kept by two breeders, in addition to a few zoological gardens, when the precipitous decline was realized. David Whittaker of Milwaukee, who had previously worked in the timber and mining industries in the Upper Midwest and

Canada, had a personal aviary with passenger pigeons as "pet birds." He had acquired two pairs in 1888, although the older pair soon died. He successfully bred the younger pair in captivity, so that when ornithologist Deane visited in March 1896, there were about fifteen birds there.[58] Deane "hoped that, if Mr. Whittaker continues to successfully increase these birds, he will dispose of a pair to some of our zoological gardens, for what would be a more valuable and interesting addition than an aviary of this rapidly diminishing species."[59] Although the evidence had mounted that the pigeon was nearly extinct by 1896, Deane's response reveals that, even when facing the end, some people continue to hope that a solution will be found. This is one of the normal emotions within the grieving process.

Soon after Deane's visit, the flock was purchased by Charles Otis Whitman, although seven birds were returned to Whittaker in 1898.[60] Whitman, who became a professor of zoology at the University of Chicago in 1892, worked extensively with breeding pigeons, raising over thirty species of pigeon in coops at his home.[61] Whitman was particularly interested in scientific questions of genetic inheritance, as well as the viability of cross-species breeds. He established a pigeon breeding center in Woods Hole, Massachusetts, which he visited each summer. In March 1896, he purchased three passenger pigeons from David Whittaker, then bought another pair from Whittaker in October and finally Whittaker's remaining ten in March 1897. Four of the hatchlings in 1897 lived, giving Whitman a flock of nineteen. The following years proved less successful, with fewer eggs hatching and fewer young surviving. In 1901, five hatchings survived, although the gains were offset by the death of three adults.[62]

Whitman's early successes at passenger pigeon breeding were hailed by *Boston Evening Transcript* in 1901 as a "revival of this almost extinct species."[63] But the attempt failed: by 1903 Whit-man had less than a dozen, who were all descendants of a single pair. In 1906, Whitman's flock was a meager five individuals, "the last representatives of a species around whose disappear-ance mystery and fable will always gather."[64] None of Whitman's birds would ever be released into the wild.

As the flocks in captivity dwindled to nothing, ornithologists attempted to mobilize a dramatic search for any wild pigeons that could still be conserved. At the 1910 annual meeting of the American Ornithologists Union (AOU), Clifton Fremont Hodge, professor of biology at Clark University, presented a paper titled "The Present Status of the Passenger Pigeon Problem," in which he argued that scattered passenger pigeon sightings called for "an effort to save them."[65] A reward of $300 was offered by Col-onel A. R. Kuser for information leading to a pair of undisturbed nesting passenger pigeons, and others issued similar rewards for identification of living birds in particular states.[66] The rewards applied only to live birds because the idea was to protect the living animals and any nests. Hodge's idea was to start a "Passen-ger Pigeon Restoration Club" that would "take up the protective work, secure legislation and warden service, so the birds may breed in safety and again range the continent" if some live birds could be located.[67]

The search was on. In November 1910, Hodge reported on the "search of the American continent for this lost species."[68] In an unexpected interpretation of the reward, Hodge had begun receiving nests by post because the newspapers had carefully stated that the rewards were based on nests, not dead birds; for-tunately, these had all been mourning dove nests. There had been no evidence of passenger pigeons, but Hodge claimed that "negative evidence is proverbially inconclusive" and still held out hope that a couple of seasons would be needed to decide whether or not to abandon the search.[69] His reason for "prolonging the misery" of the search was simply that he "could not let go."[70] The grief would not loosen its grip on him. He recognized the futility of the search, suspecting that "the worst fears of American naturalists were about to be confirmed and that we are 'in at the death' of the finest race of pigeons the world has produced."[71]

The search was continued into 1911, but no confirmed passenger pigeons were found.[72] Hodge remarked that the great irony was that many of the sightings came along with reports that the birds were killed, even though the rewards specifically stipulated that the birds were not to be harmed: "The nightmare of the whole situation has been that the last survivors of this great species were being ignorantly shot off," he wrote in exasperation.[73] Despite this sentiment, Hodge reissued the rewards for 1912. None were claimed. The only good that had come out of the attempt to locate the birds was "awakening the country to the problem, and this awakening can, and doubtless will, be utilized in saving other species which are in present danger."[74] Hodge's grief shows a transition from a bargaining phase, in which hope still held out, to a state of misery and depression.

Depression: Pigeons and a Profound Sense of Loss

On September 1, 1914, in the Cincinnati zoological gardens, the last known passenger pigeon died.[75] Martha had spent her whole life in captivity. She had been one of the passenger pigeons in the flock held by Professor Whitman and was sent to the zoo in 1902.[76] She had lived alone since 1910, when she lost her mate, George (the pair had been named after the first president of the

United States, George Washington, and his wife, Martha).[77] They had never successfully produced offspring.

Although the imminent death of the entire race of passenger pigeons engendered a narrative of anger, the final passing of this one pigeon brought out a new narrative structured around loss. Martha had been in poor health for several years, and at the end of August 1914 her death was imminent. The day before her death, the *Washington Times* issued an article, reprinted in other national papers, that began like a hospital visit to a dying friend: "In the Cincinnati zoological gardens there is dying today, if the bird is not already dead by the time this appears in print, the last known passenger pigeon in the world."[78] Lamenting the loss of the last was paramount. The *Cincinnati Enquirer* announced her death on September 2 this way: "'Martha' is dead. In one great respect she resembled Chincatgook, the 'Last of the Mohicans,' for she was the last of the Passenger Pigeons."[79] Her death was a "calamity," according to a later commentator.[80] All these news articles portray the passenger pigeon's passing as an event that should be mourned by all.

When she was found dead, her body was put on ice and shipped to the Smithsonian in Washington, DC. R. W. Shufeldt performed a detailed autopsy as part of the taxidermy prepara- tions of the body. Shufeldt documented in detail the work to skin the body and provide a scientific anatomical description, taking numerous photographs along the way.[81] It appeared that Martha had suffered from severe internal bleeding, with her liver and intestines disintegrating before the autopsy. Although he attempted to write a distant, objective, scientific description of the anatomical features of passenger pigeons, Shufeldt was moved by his work on this particular bird. His description of his work when he got to the heart betrays his true feelings: "I

therefore did not further dissect the heart, preferring to preserve it in its entirety,—perhaps somewhat influenced by sen-
timental reasons, as the heart of the last "Blue Pigeon" that the
world will ever see alive. With the final throb of that heart, still
another bird became extinct for all time,—the last representative
of countless millions and unnumbered generations of its kind
practically exterminated through man's agency."[82] The profound
sense of loss felt by Shufeldt is evident here in an otherwise scientifically neutral text.

In 1932, the Smithsonian issued a report declaring the formal
extinction of the bird and put the mounted body of Martha on
display. Martha's skin was made into a taxidermy mount and
was displayed in the National Museum's Bird Hall, then moved
to the Birds of the World exhibit. Martha on display at the Smithsonian was "a sad reminder of the once great glory of her vanished race."[83] The Smithsonian report noted frequent claims of
passenger pigeon sightings, but these were always either mourning doves or band-tailed pigeons. The report dismissed theories
that continued to pop up claiming the birds had either relocated
south or were all wiped out in a catastrophic weather event.[84]

Just as the loss of the Swedish beaver had brought about a
communal sense of guilt, the extinction of the passenger pigeon
led to communal grief. Loss was poignantly foregrounded in
the creation of memorials and monuments to the extinct passenger pigeon in the post–World War II period. These monuments became physical manifestations of the inward emotions
of their sponsors, which are visible not only in the monuments
themselves, but also in the discourse that surrounded their
establishment.

Two memorials were dedicated in 1947. A group of boy
scouts sponsored the first memorial in the Pigeon Hills near

Abbottstown, Pennsylvania. The marker was an eight-foot stone shaft with a granite plaque, topped with a bronze passenger pigeon statue.[85] According to the plaque, the monument was placed in an area known as Pigeon Hills near Harrisburg, where the passenger pigeon had flocked "from earliest pioneer days until the 1800's [sic]" and "was once so plentiful its numbers darkened the skies."[86] The scouts set up the monument "in the interest of the preservation of wild life," as an act of remembrance for the "ill-fated" bird. The monument was later relocated to a state park.[87]

The other 1947 commemoration was a stone monument with a bronze plaque set up by the Wisconsin Society for Ornithology in Wyalusing State Park (figure 4.2). This monument became widely known because the famous conservationist Aldo Leopold wrote a text about it, titled "On a Monument to the Pigeon." Leopold's text was first delivered orally at the annual meeting of the Wisconsin Society for Ornithology in April 1946, then printed in a book, *Silent Wings*, which was published on the occasion of the monument dedication ceremony, and subsequently included in his *Sand County Almanac*.[88] Leopold framed the gathering as a funeral: "We meet here to commemorate the death of a species. This monument symbolizes our sorrow. We grieve because no living man will see again the onrushing phalanx of victorious birds, sweeping a path for spring across the March skies, chasing the defeated winter from all the woods and prairies of Wisconsin. ... We, who have lost our pigeons, mourn the loss."[89] Leopold highlighted the grief he felt (and, by extension, that we readers should feel) for the passenger pigeon's extinction. The loss of the pigeon led to profound sorrow, which also activated nostalgia for the time when the pigeons were plentiful. Leopold looked back to the time when there was "the glory of

Figure 4.2
Passenger pigeon monument in Wyalusing State Park, Wisconsin. *Source:*
Author, 2017.

the fluttering hosts" and "the feathered tempest" as a time that
was alive like a "biological storm." His reminiscence, however,
is just as much about a human way of life as it is about the lost
pigeons. Leopold says that those who directly killed the pigeons
("our grandfathers") were only acting on behalf of members of
the current generation, who believe it is "more important to
multiply people and comforts than to cherish the beauty of the
land in which they live." Leopold closed his eulogy with a plea
for the "love of free sky, and a will to ply our wings" as a way for
humans to move through their grief.

Loss and grief for the pigeon were on Leopold's mind. But his
grieving was not for specific dead pigeons, individuals who had
been trapped or shot or swatted down with brooms from their

perches. These were not who the monument was for. Rather, the monument was for the loss of a species, a wholly intangible yet remarkably tangible thing that could never be replaced, or so Leopold thought.

The *Silent Wings* dedication text written by Walter E. Scott made it clear that the monument was a dedication to the species, rather than a single individual. The monument should "make us stop our busy life a moment for solemn thought." Scott urged readers to move through their grief and "lend our efforts in the direction of the perpetuation of our native wildlife in its native habitat."[90] The same sentiment was repeated in other essays in *Silent Wings*. Hartley H. T. Jackson's piece "Attitude in Conservation," for example, derided an editorial published in the *San Antonio News* in January 1947 for saying that we should not lament the "passing of a non-essential animal"; instead, he argued, the monument to the last Wisconsin passenger pigeon was not only "a token to the dead and the past, but rather as a symbol to the living."[91] The grief, then, should not make us become melancholic and inward-looking, but rather spur on change for the future.

In 1976, the US government issued a pamphlet titled "A Passing in Cincinnati, September 1, 1914" to commemorate the passenger pigeon's extinction.[92] Like Leopold's text, the pamphlet is set up as a funerary commemoration for Martha, including a graphic style that makes each page look like a gravestone. The text announcing the death of Martha from the *Cincinnati Enquirer* starts the work, followed by her portrait. After retelling the species extinction story, the text returns to Martha herself and her body as monument. Twice she flew after death—once in 1966, when she was sent to the San Diego Zoological Society's Golden Jubilee Conservation Conference, and again in 1974, when she appeared in Cincinnati as part of the launch of the Passenger Pigeon Memorial Fund to restore the birdhouse where she (as well as the last Carolina parakeet, who is often forgotten) died.[93] The pamphlet is a text written as grieving at a graveside. "A Passing in Cincinnati" assumes that grief for the pigeon will be permanent, just as Martha would be on permanent display.

Martha was not, however, kept on display permanently at the Smithsonian. Her body was put into storage in 1999 when the Birds of the World exhibit was shut down to make room for a new Hall of Mammals.[94] She would not appear again in the public eye until 2014, to mark the centenary of her death and the extinction of the passenger pigeon. By the time she made her centenary appearance, talk had already begun about her not being the last.

Acceptance? De-extinction and Longing for Passenger Pigeon 2.0

The February 2012 Long Now Foundation meeting at Harvard took place amid these stories of the passenger pigeon that had passed from this world. Building on the grief narrative, the participants in the meeting were instigating a new story: one that shooed away acceptance of the loss and returned to a bargaining phase. The loss of the pigeon had led to a longing for bringing the pigeon back, and this was a longing that might be possible to realize through new techniques. Maybe Leopold would have thought the same if the technology had been available at his fingertips, or maybe he would have been content with remembering and moving on to the final acceptance of the pigeon's demise. But the people in the room on that winter day were not content to accept that the birds that had once numbered in the billions were gone. They wanted to move forward by turning back the clock.

The meeting launched the Revive & Restore project of the Long Now Foundation. The project rapidly expanded its reach with a second meeting in October 2012, which tripled the number of participants, and then the project hit a high point with a TEDxDeExtinction event hosted by National Geographic in Washington, DC, in March 2013. National Geographic featured de-extinction as the cover story of its monthly magazine, and the TEDx event was streamed live on the internet.[95] The program was chock-full of de-extinction proponents, who discussed the potential de-extinction candidate species, the technical possibilities of genetic technologies, and histories of conservation efforts that brought nearly extinct species out of danger. There were presentations spouting words of caution, but these were few and far between.

Longing for lost species was a common theme in many of the presentations at the TEDx event. For example, the artist Isabella Kirkland showed her painting *Gone*, characterizing it as a way to capture "human-motivated loss in people's minds over time." Mike Archer, a professor at the University of New South Wales, argued that "we've got a moral imperative to try to do something if we can" in the case of an extinction caused by humans. According to Archer, resurrecting a species would "restore the balance of nature that we've upset" and "put it [the species] back where it belongs."[96] The longing for and belonging of extinct species in the twenty-first century was the overarching message.

Passenger pigeons turned into a de-extinction poster child, along with the mammoth. Two of the talks at the TEDx event were specifically on the passenger pigeon, and the species featured prominently in the National Geographic coverage. Press coverage of the event often reproduced drawings of passenger pigeons, such as Audubon's watercolor, and highlighted the passenger pigeon case.[97] From the perspective of the Long Now Foundation, the passenger pigeon had always been at the center of de-extinction and the related new subdiscipline called *resurrection biology*.[98]

"The Great Passenger Pigeon Comeback," as that part of Revive & Restore came to be known, is called its "flagship project." Ben Novak, who is the scientist leading the work in the Revive & Restore project, has stated that bringing back the passenger pigeon is "about the pigeon's place in the forests of tomorrow. To know how to proceed we must explore the past. My thoughts in the pursuit of de-extinction are to the great flocks of passenger pigeons and how they shaped the forest."[99] The millions of birds certainly would have been significant distributors of seeds and consumers of forest resources. In his view, conservation is at "a turning point that the de-extinction of the passenger pigeon can sway towards the value of life, just as the extinction of the passenger pigeon did so powerfully a century ago."[100] That flip of conservation is something that Stewart Brand has argued is a primary reason to use genetics to bring extinct species back to life: "That something as irreversible and final as extinction might be reversed is a stunning realization. The imagination soars. Just the thought of mammoths and passenger pigeons alive again invokes the awe and wonder that drives all conservation at its deepest level."[101]

This is a radically different framing for recovering the passenger pigeon than experienced earlier in the story of its loss. The emotional basis for the intervention was still grief at the loss of the pigeon, but rather than accepting the passenger pigeon's fate as already made, the Revive & Restore practitioners have essentially entered a new bargaining phase based on hope and wonder. The idea that the grief can be ameliorated by righting the wrong of the extinction is driving de-extinction forward. The project is currently in the in vitro phase while ongoing in situ research continues.[102] The Revive & Restore project's roadmap has an ex situ phase beginning in 2022 and soft releases of passenger pigeon 2.0 individuals into the wild beginning in 2027.

Revive & Restore may, however, have entered a devil's bargain in its attempt to resurrect the passenger pigeon as a mode of dealing with collective grief. First, while the project may end up with an animal to release, there is little reason to believe that the pigeon will be successful in recolonizing the Midwest. Most of the land has been modified beyond recognition by agriculture, and those humans that inhabit the bird's original territory would not likely share the land. Even in 1970 on the fifty-sixth anniversary of the passenger pigeon's extinction, Texas columnist Marjorie Adams, who had a syndicated newspaper series called "Bird World," commented: "We of today decry the slaughter that brought the extinction of the elegant and tasty passenger pigeon. Yet if we had the pigeon returned to us is anyone foolhardy enough to believe that agricultural American would tolerate its existence in such numbers for long?"[103] Second, there is the potential that the social meaning and significance of the passenger pigeon resurrection will focus on the fantastical, science-fiction qualities of the project rather than the biological or ecological science and its implications. This is precisely what happened in the Thylacine Cloning Project of the Australian Museum from 1999 to 2005.[104] Few, if any, journalists can resist the reference to *Jurassic Park* when covering a resurrection biology story. The fantastical framing may negate any potential power of the grief narrative and yet still not take advantage of the comedic narrative that Heise advocated.[105]

The 2012 workshop on the resurrection of the passenger pigeon had been aptly titled "Forward to the Past." The new genetic technology was the forward-looking aspect, whereas the longing for the return of the bird that motivated so much of the work was grounded in the past. The grief of losing this species that had once numbered in the billions has permeated the discourse from Hornaday to Brand. Humans commenting on the loss of the passenger pigeon do so time and time again using the emotional framework of grief. They believe that this bird lost through extinction belonged fundamentally to the American landscape. The Revive & Restore project promises that perhaps the lost bird may again be found through its resurrection.

5 Remembering: Narrating Species Loss and Recovery

Bruno stands tall as a problem bear: knocking over a cart, grabbing a handful of honey to stuff into his already full mouth (figure 5.1). He appears unruly and wild, yet strangely anthropomorphized as he stands semiupright among farming technology like a child caught with a hand in the cookie jar. He is nature that has invaded the human world.[1]

Born in an Italian nature reserve, the bear known to science as JJ1 was an ordinary brown bear. He was regularly monitored by scientists interested in recovering brown bears in Italy until May 2006. At that time, he wandered across the Italian-Austrian border and was spotted on May 20 in German Bavaria. With this wandering, he became the first brown bear on German soil since the species had been exterminated 170 years ago. He had single-handedly reintroduced his species to part of their former range. But it turned out to be an unwelcome reintroduction. In the course of his transgressive venture, JJ1, who became known as Bruno in the German media, supposedly killed thirty-three sheep, four rabbits, and a guinea pig, as well as raided a honey-bee farm. A group of Finnish hunters was contracted to snare the bear for relocation, but in the end he was shot by other hunters. Bruno's life came to an end on June 26, 2006, less than two

Figure 5.1
Bruno the Bear on display at the Museum of Man and Nature, Munich, Germany. *Source:* Author, 2018.

months after moving over a political boundary unbeknownst to him.

Bruno's taxidermy-prepared body was put on display in the Museum of Man and Nature (Museum Mensch und Natur) in Munich, where I encountered him in 2013 and again in 2018.

Through that display, the museum is participating in the process of remembering not only the life and death of this particular bear, but the losing and finding and losing again of brown bears in Germany as a whole.

The processes of recovering lost nature are built upon things remembered. To identify something as lost *now* requires knowing (or believing) it was there *before*. How we remember the presence or absence of animals in the landscape and our lives, whether they are specific individuals like Bruno the Bear or whole populations like beavers in Sweden, shapes our emotional responses to them. Remembering the past in the present is how we decide that something *belongs* and that we *long* for its return. The memories, however, need not be personal; they can be passed on through collective memory processes.

The term *collective memory*, as seminally proposed by Maurice Halbwachs, represents how collective experience beyond the individual becomes remembered as it is transmitted through narratives, whether oral, written, or visual, and in many settings, including literature, art, ritual, and the media.[2] Within memory studies, which is a vast and sprawling research field, a distinction has been made between two kinds of collective memory: *cultural memory*, which is institutionalized, recounted by specialists, and requires external symbols to invoke the memory; and *communicative memory*, which is noninstitutional, based in everyday action, and has a limited time depth of three to four generations.[3]

The distinction, however, is strictly enforceable only on the theoretical level because of the linking of memory practices. As Harald Welzer points out, cultural memory over the long term can shift because daily practices of communicative memory de/revalue parts of the story.[4] More than just transmitting narrative facts, communicative memory involves creating organizational structure and cultural connotation of those facts, which allows stories to be retold in adapted form to the needs of the present.[5] As such, communicative memory as deployed in a certain place and a certain time reflects both the past and the present. Events occur and are then translated and transmitted over time, making memory of events key to understanding them. An approach to history that focuses on memory-making and memory-deploying, or *mnemohistory*, as coined by leading memory scholar Jan Assmann, is "concerned not with the past as such, but only with the past as it is remembered."[6] The practices of communicative memory-making and memory-deploying affect what we know about things.

Memory affects the understandings people have of their environments. Cultures create associations with the landscape over generations, while at the same time individuals connect personally with it.[7] Fisheries scientist Daniel Pauley launched the idea of *shifting baseline syndrome* into scientific discourse in 1995. In his work, he argued that fisheries scientists and managers were guilty of basing calculations on degraded ecosystem numbers rather than pristine ones: "Essentially, this syndrome has arisen because each generation of fisheries scientists accepts as a baseline the stock size and species composition that occurred at the beginning of their careers, and uses this to evaluate changes. When the next generation starts its career, the stocks have further declined, but it is the stocks at that time that serve as a new baseline."[8] The shifting baseline syndrome has been applied in conservation science to discuss things such as perceptions of ecological change, how to plan for ecological restoration, and integration or rejection of newly introduced (or reintroduced) species.[9] People tend to view the environment of their childhood as the "right" one, and thus things that changed before that are imperceptible unless they hear stories told about the way it used to be. In other words, shifting baseline syndrome is about memory practices: what's remembered and what's forgotten.

Collective memory is also dependent on the histories told—the explanations given of how things were in the past and why they are the way they are now. History (whether it happens in formal history writing or other history-creating enterprises, such as monuments and museums) reinterprets the past through the present, which in turn changes how we remember the past.[10] Of course, memories are not just about the past, but also about imagining futures.

Museums feature in many of the remembrance places I discuss in this chapter. That is because they have the ability to construct "plots" that bridge the gap between the scientific and the affective, to arouse emotions in the public; exhibits can incorporate private memory to develop a shared public experience.[11] Natural history exhibits, like the one featuring Bruno, have long been designed to impart a moral education as much as a natural history one.[12] Natural history museums often have an explicit mission to advance public understanding of the natural world and our human history in it, as well as instill an environmentally friendly ethos.[13]

In this chapter, I will argue that memory can have a profound effect on notions of longing and belonging. The animal-recovery projects discussed in this book have been driven by the history of the relationship between humans and the species, as well as the ways in which that history is remembered. Deploying collective memory as a framework allows environmental historians to think about how relayed historical narratives change the way people think about, feel about, and potentially act to address environmental issues in the present. I'll begin with Bruno's story and then circle back to each of the recovery stories I've covered in this book.

Communicating Bruno's Life and Death

European brown bears (*Ursus arctos*) have been on the verge of extinction in the Alps for the last century. Although the Italian government issued a total ban on bear hunting in 1939, bears were found only in the Adamello Brenta region in northeastern Italy by World War II, and by that time there were no more than fifteen individuals.[14] Even with legal protection, the population failed to recover, and by the mid-1990s there were only three bears left. To stave off total extinction of the brown bear in Italy, a reintroduction program was launched in 1999 to relocate bears from Slovenia, where they had found reproductive success, to the Italian Alps. Ten bears were released between 1999 and 2002 into the Adamello Brenta nature reserve by the EU-funded LIFE project Ursus.[15] Bruno's parents, Jurka and Joze, were both bears reintroduced into Italy from Slovenia.

Bruno was born in Adamello Brenta in 2004, and when he reached the age of two, he ventured away from his mother's protection. But he journeyed a bit farther than people had projected when they reintroduced the brown bears into Italy: on May 5, 2006, he was spotted in Austria. Throughout the month, he moved large distances, finally making it into Germany two weeks later. For the next month, he continued to roam back and forth over the Austria-Germany border, all the while raiding farms for food. Because of these behaviors, he was categorized as a "problem bear" under the Austrian brown bear management plan and needed to be relocated or killed.[16] Although attempts were made to find him so that he could be drugged and relocated, in the end, he was shot. The whole saga received a huge amount of publicity and regular news coverage, partly because Bruno was adopted as an informal mascot for the 2006 World Cup, which was being hosted in Germany at the time, so many were outraged by the final result.[17] In the media coverage of the saga, it was pointed out that Bruno was the first bear in Bavaria for 170 years.[18] The brown bear was lost from Germany, briefly found, and then lost once again.

Bruno's story was thus one about the recovery of a species that didn't happen—at least, not in Germany. Bears are one of the animal species being targeted by the Rewilding Europe foundation in its programs to conserve and expand wilderness in Europe, with programs in the Central Apennine mountains and Croatia.[19] The established populations are expected to disperse northward with contemporary climate change, so Bruno's incursions should have come with little surprise. The motto of Rewilding Europe is Making Europe a Wilder Place, something that this young male brown bear was doing naturally. The human response to Bruno's (mis)behavior reveals the limitations of rewilding initiatives: we may like the nostalgic idea of free-roaming wildlife, but only so long as it doesn't interfere with our own borders and cultural norms.[20]

How has Bruno's story, a story about the unsuccessful self-reintroduction of brown bears to Germany, been remembered? In the Museum of Man and Nature, Bruno himself is displayed as a destructive bear, yet the exhibit also puts a question mark after "problem bear." This isn't a straightforward narrative. On one side of the exhibit, there is a minishrine to the bear, complete with a picture of the marked crosses set up where he was killed, a flag, a t-shirt, and a commemorative "Brown bear Bruno JJ1" toy bear made by the famous teddy bear company Steiff (figure 5.2).

Next to this makeshift shrine, the curators have placed a wall on one side of the room on which visitors can write their own messages and thoughts. When I visited in 2013, many messages expressed guilt or grief for Bruno's death, along with hope that the same thing would not happen to future bears that crossed the invisible line on a map.

Figure 5.2
Portion of the JJ1/Bruno exhibit at the Museum of Man and Nature, Munich, Germany. *Source:* Author, 2018.

Putting a natural history specimen on display can, in fact, serve to allay the guilt for the death of the animal, or even the species. Hanna Rose Shell has argued that when William T. Hornaday intentionally killed American bison that he knew were sliding dangerously close to extinction, he did so with the idea that putting the taxidermy versions of those animals on display would be "atoning in small part for the devasting slaughter of the buffalo that had occurred over the preceding century."[21] Taxidermied specimens can memorialize the losses as "a gesture of remembrance," as Rachel Poliquin has put it.[22] The curatorial choices in the Bruno the Bear exhibit ask the viewer to remember the bear as both a physical animal and an embodiment of nature's unfulfilled recovery.

Immediately after Bruno's death, film specialist Petra Fohrmann published a small format book, *Bruno alias JJ1: Reisetagebuch eines Bären*, to record Bruno's journey. This book about Bruno is available for purchase in the Museum of Man and Nature gift shop. Fohrmann imaginatively paired real press reports about the "problem bear" on the left-hand pages with a fictional first-person narrative from Bruno's perspective on the right. Whereas the media was concerned with how many sheep he has killed or how many years it had been since a brown bear had been in Germany, Bruno is thinking about the great opportunities to find food in the area.[23] While newspapers are reporting plans to track down the bear and relocate it, Bruno is contemplating finding a mate and settling down.[24] When the press laments problems finding Bruno's tracks, the bear is happy not to hear barking dogs nearby.[25] The narrative construction asks the reader to imagine being the bear. That imagining comes with a sense of belonging and emotional happiness: Bruno doesn't know he is in a place that humans say he doesn't belong in, and he has no idea that the humans are unhappy about him being there.

A similar approach was taken by Heinz Vogel in his picture book on the adventurous life of Bruno published in 2012.[26] The book's forward lays out the "real" story of Bruno, ending in his death, but Vogel asks the readers to imagine a story with a different ending. Told again from the bear's perspective, the narrative has little Bruno learning how to be a bear and then striking off on his own. The hunter sets a trap for Bruno, but instead of Bruno dying, the hunter and his dog do. Bruno then emigrates to Canada where bears are allowed to be bears. Despite the tragic ending to Bruno's life story, Vogel's book offers a counternarrative of hope for recovery: if the bear had been welcomed like it would have been in Canada, then Germany would have had his delightful bear cubs and happy bear family.

Bruno's end has been marked in memorial and memory. In addition to crosses set up at the site of Bruno's death in Germany, two monuments in Austria have been erected in his honor: one at the alpine lodge Gartalm near the town of Zillertal, and another at the Lamb Hütta lodge in the village of Gaschurn-Partenen.[27] When the tenth anniversary of the Bruno saga arrived in May 2016, many regional newspapers ran articles of remembrance, but some pointed toward the future. A contribution in the *Augsburger Allgemeine* asked "can a new bear arrive?" while another in the *Tiroler Tageszeitung* categorized the conflict over the return of wild animals as "greater than ever."[28] Bruno's untimely death has been marked, but little significant change to policies about brown bear reintroductions has resulted. There are still no brown bears in Germany and no concrete plans to bring them back.

In the case of Bruno, the public outcries of grief—from the crosses and monuments to t-shirts and teddy bears—are personal. The Bruno remembrances name this individual bear, rather than all brown bears. They fail to meaningfully engage with the larger issue of bear reintroduction in Europe and the myriad of feelings—hope, fear, grief, and guilt—that can be tied to it. The narrative emphasizes the loss of a particular bear who can never be found again.

An Anniversary of Death

Like Bruno's personal story as a story of grief, the story of Martha the last passenger pigeon came to the fore when the centennial anniversary of the passenger pigeon's extinction came around in 2014. The stuffed and mounted skin of Martha, who had been put into storage in 1999, made an appearance in an exhibit called *Once There Were Billions: Vanished Birds of North America*, curated by the Smithsonian Library. The exhibit was on display from June 24, 2014, to January 3, 2016, in the Natural History Museum. The small exhibit, which was tucked under a stairway on the bottom floor, featured two cabinets with four extinct North American bird species: heath hen (extinct 1932), passenger pigeon, Carolina parakeet (extinct 1918), and great auk (extinct mid-1800s). Each species display included at least one mounted, taxidermied specimen and book illustrations (mostly originals). Martha was the only "named" bird in the exhibit text, although we know that the last heath hen and Carolina parakeet were also in captivity and were referred to by anthropomorphic names. The exhibit signage claimed that the stories of the extinct birds "reveal the fragile connections between species and their environment."

In this public appearance after her death, Martha sat on a branch with her body facing away from me, her head turning in my direction. I stared at her as she seemingly stared back with a glassy red eye. In a way, she looked too lifelike to sympathize with, to wonder what she had thought about as she spent her last few years without her mate, George. Now she was placed near another potential mate. A male bird, one Martha never knew, reached out with a seed in beak, possibly as an offering to Martha's passing. But he too seemed too real, too alive. The third bird, laying down, did not: in front of Martha and the male, a passenger pigeon skin specimen lay red belly up and had no eyes. This skin embodied the death of the passenger pigeon. Behind the birds, images of passenger pigeon hunting were reproduced on a grand scale. The images seemed to suggest that Martha and the male would be the next to die. There was no escaping the extinction. It was fact. After the exhibit closed, Martha as specimen was interred once again in a storage room.

As part of the Smithsonian exhibit, a passenger pigeon sculpture was placed outside of the Museum of Natural History in a small garden space. The large-scale black passenger pigeon twists its neck upward, perhaps in a posture of scanning the trees for comrades who will never be seen again. Artist Todd McGrain sculpted the passenger pigeon bronze statue and four others in his Lost Birds series "to create shapes that capture the presence of the birds, to make them personal and palpable, to remind us of their absence."[29] Although the Smithsonian version of the statue was not a permanent fixture, another copy of McGrain's passenger pigeon statue still sits at the Audubon Center in Columbus, Ohio, which is one of the last areas where passenger pigeons lived. The dedication at its base reads, "In Memory of the Passenger Pigeon / Driven to Extinction 1914." Although the monumentalization of remembrance for the passenger pigeon had history back to the 1947 memorials in Wisconsin and Pennsylvania, the one-hundredth anniversary of Martha's death sparked a new wave of interest in publicly remembering the loss of the bird through art in the Midwest states.[30]

A mural painted in 2013 on the side of a four-story building in downtown Cincinnati shows a flock of passenger pigeons swooping through the air, following the lead of one pigeon out front, whom the artist identified as Martha (figure 5.3).[31] The birds fly into the sky past the pavilions of the Cincinnati zoo where Martha met her end. There is a religious quality to the image as the birds circle to freedom on a cold, snowy winter day. The swirling birds are aesthetically reminiscent of Lewis Luman Cross's *Bird's Eye View of Passenger Pigeons Nesting* (1934) in the Grand Rapids Art Museum. In that painting, a flock flies across a golden sky at sunset to land in the foreground trees and compete for a nesting spot among the crowded branches. But rather than a flight to freedom, Cross's birds have come to meet their

Figure 5.3
Martha, the Last Passenger Pigeon mural, by artist John A. Ruthven, Cincinnati, Ohio. *Source:* Author, 2017.

end: hunters standing under the trees use long sticks to knock the unsuspecting birds into sacks, which are carried off in wagons, and hunters with rifles take shots at the birds from a canoe. Whereas Cross's painting speaks to the human guilt over the pigeon's extinction, the new mural invokes hope and freedom. Although the plaque on the mural proposes that the artwork "serves as a daily reminder to downtown residents, workers, and visitors of the importance of wildlife conservation and the reality of extinction," the connection between human action and extinction is not visually present. Instead, the aesthetics focus on hope.

For its one-hundredth anniversary, the Cincinnati Zoo refurbished its Passenger Pigeon Memorial building pictured in the mural. The inside exhibition was rebranded as "Martha's Legacy: Lessons from the Passenger Pigeon for a Sustainable Future." The history of the passenger pigeon decline culminating in the death of Martha in that very building is paired with a display on wildlife conservation successes and the roles of zoos in conservation. A sculpture of a pair of nesting passenger pigeons (taxidermy specimens were too rare) give an impression of the birds in life, while nets on display show how they died.

The ceiling over the Martha's Legacy entrance sign is filled with paper birds from the Fold the Flock initiative (foldtheflock .org), a passenger pigeon origami project. The project developed an origami template for a passenger pigeon and made it freely available for download. Individuals and groups were encouraged to contribute to the flock on the participants page, and as of April 2016, the website claimed that 1,359,590 origami birds had been folded for the flock. Like the mural, this project is about putting the birds into flight. Although there is an educational opportunity to learn about the passenger pigeon extinction, the emphasis was on having a creative experience with the long-dead pigeon.

The narratives of the 2014 commemoration events tended to focus on learning a lesson from the passenger pigeon extinction so that the same mistake would not be made again. Exhibits were staged in museums throughout North America and abroad, highlighting their passenger pigeon specimens in special cases in 2014. The Museum of Natural History at the University of Michigan, for example, put on the exhibit *A Shadow over the Earth: The Life and Death of the Passenger Pigeon*, which used "the story of the passenger pigeon as a cautionary tale and a call to action." *Final Flight* at the Harvard Museum of Natural History framed the exhibit as an opportunity to "see one of the world's last mounted specimens of this now vanquished bird and learn how its extinction inspired the protection of other species." The exhibits stressed learning from the past, which is taken as an accepted (albeit not pleasant) part of history.

Some popular science books published to coincide with the centennial—Joel Greenberg's *A Feathered River across the Sky* (2014), Mark Avery's *A Message from Martha* (2014), and Errol Fuller's *The Passenger Pigeon* (2015)—set their stories up as ones with a message of moving toward a more positive future. The tenor of these works was different than the older narratives of loss. Now the story, as Greenberg put it, was being told with "a broader hope that this centenary could be a vehicle for informing the public about the bird and the importance that its story has to current conservation issues. ... Our goal is to use the centenary as a teaching moment."[32] In his introduction, Fuller likewise hoped his book "will help to bring awareness to just how fragile the natural world can be."[33] The teaching moment through the passing of the passenger pigeon out of the world turned the narrative toward education rather than emotion. Avery was motivated to write about the passenger pigeon because "extinction is general deemed to be a 'bad thing' … and, if we believe it is a bad thing, then we should do more to prevent future extinctions."[34] These books express remorse for the extinction, but they also observe a desire to move on. The question becomes: Where to move on to?

Memories of Muskoxen

Part of moving on involves recognizing the past you're moving away from. This proves to be a challenge in the case of the muskox. The muskoxen brought to Norway in the twentieth century are very recent additions to the Scandinavian environment in ecological terms, yet that doesn't mean that people remember a time before muskox. Because the animals were first brought to Norway in the 1930s, they have always been in Scandinavia within living memory. Because the animals have made their homes here for more than eighty years, collective memory plays a part in whether the muskox is framed as belonging here.

When I went on a muskox safari in Dombås, Norway, in 2013, muskox memory came to the fore. On my four-hour tour to see the wild muskox of the Dovre mountains, I had the pleasure of hiking with Heidi, a lawyer from Oslo, and Saul, a corporate coach from London, along with our guide, Joakim, a twenty-five-year-old ecology student born and raised in Dombås. What I noticed was the way that memory of recovered animals works—or does not.

Near the beginning of my tour, Joakim stopped us to talk about what to do if we ran into a muskox close up. He was very intent on stressing the calm, peaceful nature of the animals;

they would only attack if threatened, he said, so we had nothing to worry about. As a historian, I have read a lot of newspaper accounts of muskox attacks, and though he is right that they attack only when threatened, I wouldn't characterize them as peaceful; their response to a threat is often to attack. When I mentioned to him that a man had been killed by a muskox only a few kilometers up the road in 1964, our guide was shocked. He had no idea that someone had died from a muskox attack in the area. I was a bit surprised that the death hadn't entered local lore, especially because it prompted a fair amount of community outrage, but that history obviously hadn't trickled down to this young ecologist. Without this history, Joakim could think of the muskox as a different kind of creature than he would with it.

At one point, I asked Heidi what she as a Norwegian thought about the muskox. Her answer was that the muskox had been in Norway as long as she had been alive, so they were a normal, natural part of the Norwegian countryside. Although she had never seen the wild muskox herd before, she had never thought of them as out of place. This is a typical way of thinking about the past. For many, if something has been present for his/her lifetime, it becomes the baseline, as described by Pauly. Humans have a lifespan that, compared to ecological and even historical timeframes, is a blink of an eye, yet we have a tendency to judge what happens by what we see with our own eyes.

Muskox as symbol has been integrated thoroughly into the Dovre area. Moskus Grillen in Dombås serves Norwegian café food and pizza under a wall-sized glass-printed muskox image; numerous outfitters take tourists on muskox safaris; and the historic Kongsvold Fjeldstue hotel offers muskox meat appetizers with dinner.[35] Tourists can buy sweaters and mittens made of muskox wool (qiviut) and visit a small museum exhibit about

Dovrefjell with stuffed muskoxen and skulls on the walls. The commune of Dovre adopted a communal shield in 1986 with a muskox and installed a bronze statue of a muskox in Dombås in 2007.[36] The muskoxen are presented generally as animals who have always been here: natural wonders to witness rather than subjects with a history.

Histories of animal recovery can also be remembered wrongly. In the Frösö Zoo's Nature Room in Östersund, Sweden, a muskox stands in the last section of the exhibit in the area labeled *Svalbard*. A sign upon entering notes that the exhibit was set up as a journey through Sweden from "Skåne to Svalbard"—which is a nice alliteration, but somewhat ironic because Skåne and Middle Sweden are in Sweden, whereas Svalbard is an arctic archipelago claimed by Norway. What is exceptional about the muskox on display is that the exhibit was opened in 1986, and the last known sighting of muskox on Svalbard was in 1985. The first introduction of muskox on the Norwegian-claimed archipelago of Svalbard had taken place in 1929 with the release of seventeen calves captured in Greenland.[37] The herd appeared to thrive: according to reports in 1936, the herd had grown to thirty, and by the mid-1960s, there were anywhere from fifty to one hundred. The population declined suddenly in the 1970s, and the whole group died out by 1985. At the same time, the breakaway herd of muskox from Norway had come over to Härjedalen (the adjacent Swedish county) in 1971 and stayed, as I discussed in chapter 3. So in 1985, there were actually muskox in the neighborhood of the zoo, but none on Svalbard.

Thus when the exhibit opened, this muskox was placed in exactly the wrong place. Although muskox had been reintroduced to Svalbard, they were no longer there—so did the muskox belong there? At the same time, live muskox were currently inhabiting Middle Sweden, so did it belong in that part of the exhibit instead? Clearly the muskox was not understood as a Swedish animal by Rune Netterström, who set up the exhibit.

Animal exhibits in museums are relatively static encounters with nature, yet nature is not static. Animals move, as the muskoxen did when they ventured into Sweden. Where do we place animals whose geographies have changed, whether by our doing or theirs? In the Biologiska Museet in Stockholm, muskox appeared only as part of the Greenland diorama rather than in the Swedish landscape because they were not in Sweden at the time the museum was designed.[38] When I saw it in 2012, nothing had been changed since that initial design, even though the animal's range had changed. Exhibits like these declare that an animal belongs in a certain place, whether or not that reflects reality.[39]

In 2013, a near disaster for the Norwegian muskox herd prompted a new dynamic view of nature and a renewed remembrance. A lung inflammation disease appeared in the muskox herds: by August 2012, it had killed forty-three animals, and by the end of the year, about 17 percent of the three-hundred-head population had died from the illness.[40] The Science Museum run by the Norwegian University of Science and Technology in Trondheim installed a special display in the entrance hall in 2013 in response to the outbreak.[41] The visitor entered the space facing the mounted adult muskox displayed on an artificial rocky outcrop in front of a large panoramic photograph of the Dovre mountain landscape. An informational board and leaflets to take home described the "powerful" muskox, "which is adapted to the dry and cold arctic climate." It lived in Europe (including Norway) during the last Ice Age, although the animals now present in Norway are "set-out animals taken from Greenland." The crisis was described in a large headline—"Great deaths of muskox in Norway in 2012"—followed by some details of the losses of the animals recorded by the scientific community. In this temporary exhibit, the muskox's history was situated in a specific place, with local meaning. The loss of the muskox in this place where it had recovered was understood as a potential tragedy.

The muskox has been so integrated into the culture around the Dovre mountains that there is little question that it is an animal that belongs there. Yet in the process, much of its history has been erased from collective memory. It becomes the prehistoric animal without a history, rather than a twentieth-century rewilding experiment. Potential lessons about hopes pinned on wild animals as markers of wildness and fears of the wild out of control are untold.

Beavering Away

Unlike muskoxen, beavers are ubiquitous in natural history museums in Scandinavia. They are often in dioramas of woodland animals, along with foxes, grouse, and even bears. Beavers are rarely the center of attention, sitting often either at the viewer's feet or tucked away in a corner, like in the Dovre National Park Center in Dombås or the Vitenskapsmuseet in Trondheim. Very often, beavers are positioned at work cutting down trees, which is probably the thing most people think of when they think of beavers. Beavers are presented in these exhibits as ordinary and ubiquitous. They don't have any stories told about them. No narratives of destruction or success grace the walls around them. If you didn't know better (and most people don't), you would assume that European beavers have always been here in great numbers.

When I mentioned during the muskox safari that I was also working on beaver reintroduction, Heidi was surprised. She replied that she had seen beavers all over in Norway and Sweden, so how could they have basically been extinct? Obviously the history of beaver—its near demise and remarkable comeback—is not something being taught in the schools or discussed by the public if this well-educated, middle-aged Norwegian knew nothing about it. To her, the current distribution of beavers must mean that they had always been here, an assumption that we know is not historically true.

This historical amnesia may happen because there isn't a pressing need (or at least people don't believe there is) to tell the beaver's history in Scandinavia. This is in stark contrast to the discourse of a contemporary beaver reintroduction project in Scotland. In 1998, Scottish Natural Heritage held a public consultation about their proposal to reintroduce beavers, which had become extinct in Scotland in the sixteenth century.[42] In 2001, the Scottish government decided not to proceed with plans because of an outbreak of foot-and-mouth disease. Another application for a reintroduction license submitted in 2005 was denied because of the presence of woodland and aquatic conservation areas in the vicinity of the release site. In 2007, a new consultation was held, a reintroduction study was issued,[43] and the Scottish government finally approved a license for release in Knapdale Forest in May 2008. The five-year trial reintroduction finally began in May 2009 with the release of three beaver families imported from Åmli, Norway—the same place that supplied the beavers for reintroduction to Sweden eighty years before.

Throughout the project, the narrative supporting the effort has stressed the corporate guilt we today should feel for the beaver's extinction. In its first public consultation summary report,

Scottish Natural Heritage included a section titled "Why should we consider restoring beavers?" This text made the case for beaver reintroduction: "There are two elements to this question. On the one hand, beavers are a missing element of our native biodiversity and were lost entirely through human activities. Many would argue that we have a moral responsibility to consider their return. Secondly, beavers are an important 'keystone' species in forest and riparian ecosystems."[44] Beaver reintroduction, while grounded in ecological principals, was presented as a righting of past wrongs. The language is one of guilt: *missing*, *lost*, and *moral responsibility*. Beavers are missing because of humans in the past and therefore humans today have the responsibility to bring them back, according to this narrative.

In 2008, when the Scottish environmental minister Mike Russell proclaimed, "Beavers were an original victim of wildlife crime when they were exterminated and it's time to rectify that," he was participating in an environmental discourse of guilt.[45] In Russell's statements, the guilt is something more than personal: beavers had been exterminated in Scotland at least five hundred years before his statements, so certainly no one listening was guilty of killing a beaver, and yet everyone was. The upshot was that we (humans in Scotland, presumably) had a moral responsibility to find and return the lost. Just as it had been in Sweden in the 1920s, guilt was called upon to encourage support for the beaver reintroduction in Scotland.

In Scandinavia now, the guilt narrative in the history of the beaver is largely absent, but this is not to say that there cannot be collective memories of these lost and found animals. As I discussed in chapter 2, the beavers were remembered in the 1920s through storytelling about the grandparents' generation or even their grandparents' time. When I went on beaver safaris in Sweden—the first in 2013 on the Dammån river near Östersund and the original release site and the second in 2015 on the Vindelven near Umeå in Västerbotten county—storytelling was still employed. While sitting on a boat, rowing for one hour out to beaver territory for a sighting on the Dammån, I heard about the history of the river and the beavers from my guide, Kurt. He began with stories of working in forestry in the early 1900s and floating timber on the river. Then he turned to the narrative of the use of beaver skin by the local populace, the beaver's extinction across Europe and Sweden, and finally the reintroduction of beavers from Norway to Sweden beginning in 1922. He set this in the larger history of beaver conservation throughout Europe and how beavers expanded their territory to this particular area. Kurt remembered when these individuals first showed up near this particular waterway about thirty-two years before my tour. His dogs reacted to something one October evening although it was too dark to see what the fuss was about. In the morning he discovered a dead beaver under a log. Now four beaver lodges are established in Dammån. But such memory is limited because it is not transportable beyond the personal interaction of guide and tourist.

An alternate route for memories is via public display, like the Bruno exhibit or muskox disease exhibit. I have discovered only two museum exhibits that discuss the beaver's recovery history in Scandinavia. The first is now lost. It was an exhibit on the nature and environment of the county in the Västerbottens museum in Umeå, Sweden, which featured a beaver right near the entrance. The beaver was presented much like in the other museums, beavering away at cutting down trees, but the text for the beaver's case told its complex story (figure 5.4). In one section, a map gave all the places in the country with names starting with *Bjur-*, the old Swedish prefix meaning *beaver*, explaining that this shows how important hunting beaver was historically. The other side gave a short history of the beaver's extinction in 1871, followed by its protection in 1873 and its reintroduction to Västerbotten country in 1924, which I discussed in chapter 2. Unfortunately, the museum underwent renovation in 2013, and all the exhibits in this section were removed. With that removal, the story of the beaver's recovery was removed from the sight of future generations in the area.

The second is at the local hunting and fishing museum Elvarheim in Åmli, Norway. This very small museum is located in the heart of beaver country—the one place that beavers still remained after their near extinction in Norway and the place from which all the beavers reintroduced into Sweden came. In addition to a large natural history section on beaver biology and lifestyle, Elvarheim has one display wall on the capture of local beavers and how they repopulated the Scandinavian peninsula. Panels include information and photographs from Robert Collett's research and local beaver trappers, as well as a map of places named after beavers (*bjor* in older Norwegian). The small rural community of Åmli has turned its beaver history into a source of pride: in 1987, a communal shield was adopted with a white beaver on a blue background. Following that decision, several local organizations and businesses adopted names and marketing material using beavers. The museum is planning a temporary exhibit on the beaver reintroduction history in conjunction with the one-hundred-year anniversary of the Norwegian capture in 1921 and Swedish release in 1922 of the first reintroduced beavers.

Both the former Västerbottens museum exhibit and the current Elvarheim one focus on when the beaver was found but

Figure 5.4
Beaver exhibit at Västerbottens museum, Umeå, Sweden. Now removed.
Source: Author, 2012.

say rather little on when and why it was lost. This means that the emotional framework which motivated the reintroducers— the deep sense of present guilt for actions of ancestors—is not present. The recovery of the beaver becomes a heroic narrative of discovery rather than a frantic search for the lost. Without an emotional framework to guide thinking about the beaver's near extinction, the modern viewer is no longer called upon to feel guilty for the history.

Emotional Frameworks of Remembrance

Circling back to Bruno, we see that the emotional framework for the brown bear recovery in the Alps is completely missing from the narrative. As a viewer, I am meant to feel some kind of guilt and remorse for the death of the bear that stands before me, frozen forever in time, but I'm not asked to feel for brown bears as a whole. I'm given no context for their species' reintroduction into Italy, which in fact explains why this bear is here. I'm left not knowing if Bruno really belonged, and even more problematically, I don't know if brown bears as a whole belong in Germany. There is a vibrant discussion of emotions people felt toward Bruno, but not toward brown bears and their absence from the landscape.

Because the Swedish beaver and the Norwegian muskox have returned to the landscape where people believed they belonged, the emotional framework of remembrance shifts: there is no longer a sense of longing for the lost because now they have been found. We can still long for the passenger pigeon because it has not (yet) returned, but that longing will fade away if it does. This has consequences. How we remember what happened affects how we interact with the here and now. As I see it, when we forget that the longing once was there, we lose a powerful part of the story; we forget what motivated people to be involved with these recovery stories in the first place. Yet understanding past motivation may be just the motivation we ourselves need to act today.

6 Reconnecting: Loving the Lost

More than anything, I hope that the histories in this book have revealed how central emotions are to the environmental restoration process. Today's scientifically minded conservationists might try to talk about how a species is integral to an ecosystem or how rewilding will reconstitute habitats in decline, but what more often than not motivates our concern with lost species is how we feel about them. This isn't a bad thing, and it's not an illogical position. Instead, it's precisely logical to look for that which you long for, the thing that belongs yet is lost.

In the beaver reintroduction history in chapter 2, I discussed how guilt can be an oppressive emotion that extends beyond individuals' personal actions to the actions of their ancestors. Feeling guilty about the beaver's local extinction permeated the actions of the Swedes who reintroduced the animal in the 1920s and 1930s. So did they do it for the beaver or themselves? The answer is really both. They believed that the beaver belonged in the Swedish countryside—it had a right, and almost an obligation, to be there. They also wanted to right the wrong of extinction that was weighing on their own consciences.

Hope of making a better future world than the world we currently live in is a foundational tenet of rewilding, as I discussed in chapter 3. There is a deep sense of longing for a past, although not with the intention of recreating the past environment so much as making a future modeled on past human-nature relations. The promoters of the muskox, both those that originally imported them to Norway and those that supported the new Swedish herd, were hopeful that the muskox would make Scandinavian nature more wild. Although the hopes of wildness pinned on the muskox are perfectly fine from afar, those who lived up close and personal with the animals experienced a wholly different emotion related to this rewilding: fear. We must not categorize fear as an unreasonable emotion; it was a perfectly reasonable and justifiable reaction to have. Whereas rewilders see hope in an animal that last roamed Scandinavia when humans were just colonizing it after the last Ice Age, others see danger.

The complicated emotion of grief permeated the history of the passenger pigeon in chapter 4, rearing its head as anger, depression, and bargaining at different stages in the process. Our present moment of a concerted attempt to create a passenger pigeon 2.0 cannot be divorced from the history of emotions swirling around the lost flocks of pigeons. The way in which the passenger pigeon narrative has developed motivates the genetic science, with grief for the loss of the billions of pigeons that once blackened the skies and our desire to find them again.

That storytelling aspect of recovery truly matters because either the extinction and recovery story becomes known by everyone (like the passenger pigeon's story) or it seems to disappear from collective memory (like the beaver's story). Choosing to focus on individuals like Bruno as empathetic vehicles or whole species like the disease-stricken muskoxen will affect the emotions of those who see or hear the story.

In emotions lie motivation. Returning to the lost beaver poster that opened this book, ATM confronts us with a call to help in its recovery. Yes, bringing it back will result in tangible rewards of abundant wetlands and more flood-resilient landscapes, but there is also the intangible reward of reconnecting with the "much missed."

Love Lost and Found

"To love that which *was* is something new under the sun": so Leopold characterized his emotions at one point in his pigeon eulogy.[1] Behind the grief, then, Leopold puts forward love as his motivator. Yet we must remember that the picture he paints of the birds' "onrushing phalanx" and "feathered tempest" were not things he had witnessed, but things he had heard about from "grandfathers, who saw the glory of fluttering hosts."[2] Is it possible that he really loved these birds that he had never personally seen?

It's easy to find love as the emotive motivator behind the reintroduction work of conservationists. I am struck by the words of an editorial penned in 2014 for the major scientific conservation-focused journal *Biological Conservation*: "At its core, conservation biology affirms that knowledge about the living world should go hand in hand with love and respect for it."[3] The scientific work of finding the lost is a labor of love. Beaver-Jensen had this kind of relationship of love for beavers. A short profile article about him in the major newspaper *Dagens Nyheter* appeared with the title "Bäverkärlek" (Beaver love).[4] Jensen was characterized as "a beaver enthusiast who believes beavers are the best of all the animals." Then the article recounted a story Jensen told the reporter about love: "I remember for example a beaver male who when he was captured was separated from his partner and youngsters. Eventually they were reunited, although in captivity. The joy of meeting again was touching. But the male was so strangely tired when he saw his beloved that he lay face down and was dead, dead from joy. Well, beaver's love is infinite, he likewise can hate to death, but never more than he has the right to." According to Jensen, the beaver loved his family so much that the joy of being reunited overwhelmed him and he died. Whether the story is true doesn't matter. The power of a love story was the point. When another, much longer profile article appeared on Jensen in the Östersund local paper in July 1936, he retold that story along with two others about the beaver's cleverness. It was one of those Romeo-and-Juliet-style sagas that was told and retold during the last few years of the Swedish beaver reintroductions.

Jensen stressed his love for beavers and their love for him on his letterhead from 1933, which featured three pictures of him with his beavers: he is holding, petting, and bottle-feeding them. When two beaver pairs were set out in Råndalen in 1935, Jensen went along with the beavers to see them safely in their new homes. His picture in the newspaper also shows him cradling one of his charges. He sent a telegram from Råndalen to Festin, who had not made the trip to because of illness, saying that the reintroduction had gone well and could be recorded by Festin, "godfather of all the Swedish beavers."[5] This is the same language that the Västerbottens läns jaktvårdsförening used when discussing their role and responsibility in caring for the newly released beavers, as I discussed in chapter 2. A sense of love, which explains why participants felt guilty for its extinction, pervaded much of the Swedish beaver reintroduction speech and actions.

Love for the living is one thing, but love for the lost is quite another. Some of the emphasis on loss, grief, and guilt when facing species extinction might be explained by E. O. Wilson's idea that humans have an instinctive bond with other living beings, a love of life or "biophilia."[6] Perhaps, however, this bond is not instinctive but learned. It is learned through the histories and collective memories shared of animals and their worlds, which is why people can be motivated to save pandas and kill wolves even if they have never seen either in person. The emotional frameworks deployed to call for beaver reintroduction or the rewilding of mountains or the resurrection of the passenger pigeon all depend on histories to instill a sense of love for something never before seen.

This love motivates the desire to find the animal again. The remnant of an animal, whether the remembrance is in physical form or in the stories humans tell each other, is not the same as the animal itself. As Aldo Leopold remarked: "There will always be pigeons in books and in museums, but these are effigies and images, dead to all hardships and to all delights. Book-pigeons cannot dive out of a cloud to make the deer run for cover, nor clap their wings in thunderous applause of mast-laden woods. They know no urge of seasons; they feel no kiss of sun, no lash of wind and weather; they live forever by not living at all."[7] Motion and emotion belong to the living, which is why the living matter. The persons behind the recovery projects discussed in this book all wanted to bring the living back to the place they once roamed, even when such attempts required strenuous effort.

Although Leopold dismissed the remnants of the pigeon as less than the real thing, those acts of remembrance did affect his own emotional attachment to the lost species and his expressions of joint societal grief for its loss. These acts have reverberated through time—with those both before and after expressing the same kind of emotions. The same holds true for the guilt felt for losing the Swedish beavers and the hope placed in Norwegian muskox. Remembering also means not forgetting. With forgetting, the animal would be framed as not belonging; there would be no longing for its return. Longing for a species who has been lost to be found is an emotional matter.

Returning to Leopold's quote—"To love that which *was* is something new under the sun"—we also find that he claims this emotion is *something new*. The upwelling of interest in bringing back species could only happen after it was recognized that they had been lost, which did not even occur to scientists as a possibility until the eighteenth century.[8] It was not until the late 1800s and early 1900s that restoration attempts surfaced, as demonstrated by all the cases in this book. These were framed as responses to human hunting and rabid capitalism that had depleted animal populations, from the beaver trapping in Sweden to the use of muskox for meat in Greenland to the large-scale killing of millions of passenger pigeons in the United States. Such concerns were not unique; complaints against the slaughter of American bison and bounties on Australian thylacine happened simultaneously.[9] The conscious acceptance that extinction was real and human-induced, and that it could potentially be reversed, awakened the host of emotions identified in this study—guilt, hope, fear, grief, and even love.

It is in this self-reflectiveness, in this mobilization of emotion to recover the lost, that the Anthropocene truly begins. Although the Anthropocene, our new age of humans reshaping Earth, has so far been thought of as a physical manifestation of planetary change, perhaps we would do well to reframe it as an emotional response to those physical changes. As the globe was rapidly transformed by increasing numbers of people, machines, and structures, and decreasing numbers of wild animals in the modern era, the historical individuals in this book wanted to counter the losses they saw around themselves. Their nostalgia for the unforgotten past—even pasts they did not experience in their lifetimes—created longing for the casualties of the Anthropocene. They decided that what belonged there is what came before that time of change and acted to recover it. The Anthropocene started when humans not only exercised their power to modify environments, but also recognized that power. The Anthropocene may be less a geological marker than a cultural one.

As the historical actors in this book identified what belonged in the landscape, they were acting in the present to bring the past into the future. The emotions that drove the recovery of lost species and landscapes were put to good use: they reconnected people to animals and environments. Their emotions tied to the lost did not create a paralyzing nostalgia, but rather a nostalgia that brings about change.

Notes

Foreword by Michael Egan

1. Richard White, "'Are You an Environmentalist or Do You Work for a Living?': Work and Nature," in *Uncommon Ground: Rethinking the Human Place in Nature*, ed. William Cronon (New York: W. W. Norton, 1996), 171–185.

2. Jan Assmann, *Moses the Egyptian: The Memory of Egypt in Western Monotheism* (Cambridge, MA: Harvard University Press, 1997), 9.

1   Recovering

1. ATM is a street artist specializing in birds threatened with extinction, although he does delve into iconic British mammals like the hedgehog and badger, too. His website is http://atmstreetart.com/. The referenced artwork is available at https://twitter.com/atmstreetart/status/47535653 5343054849.

2. Bryony Coles, *Beavers in Britain's Past* (Oxford: Oxbow Books, 2006), 185–192.

3. Carolyn Merchant, *Reinventing Eden: The Fate of Nature in Western Culture* (New York: Routledge, 2003).

4. George Perkins Marsh, *Man and Nature, or Physical Geography as Modified by Human Action* (New York: Charles Scribner, 1864), 35.

5. Marcus Hall, *Earth Repair: A Transatlantic History of Environmental Restoration* (Charlottesville: University of Virginia Press, 2005).

6. See, for example, William R. Jordan III and George Lubick, *Making Nature Whole: A History of Ecological Restoration* (Washington, DC: Island Press, 2011); and Franklin E. Court, *Pioneers of Ecological Restoration: The People and Legacy of the University of Wisconsin Arboretum* (Madison: University of Wisconsin Press, 2012). For Leopold's thinking about restoration, see Joy B. Zedler, "Ecological Restoration: The Continuing Challenge of Restoration," in *The Essential Aldo Leopold: Quotations and Commentaries*, ed. Curt Meine and Richard L. Knight, 116–126 (Madison: University of Wisconsin Press, 1999), which presents a collection of his writings on the subject.

7. Society for Ecological Restoration International Science & Policy Working Group, *The SER International Primer on Ecological Restoration* (Tucson: Society for Ecological Restoration International, 2004).

8. Society for Ecological Restoration International Science & Policy Working Group, *The SER International Primer on Ecological Restoration*, 3.

9. Marcus Hall, "Restoration and the Search for Counter-narratives," in *The Oxford Handbook for Environmental History*, ed. Andrew C. Isenberg (Oxford: Oxford University Press, 2014); Eric Higgs, *Nature by Design: People, Natural Process, and Ecological Restoration* (Cambridge, MA: MIT Press, 2003).

10. For the competing philosophical positions, see Eric Katz, "Another Look at Restoration: Technology and Artificial Nature," in *Restoring Nature: Perspectives from the Social Sciences and Humanities*, ed. Paul H. Gobster and R. Bruce Hull (Washington, DC: Island Press, 2000); and Andrew Light, "Ecological Restoration and the Culture of Nature: A Pragmatic Perspective," in the same volume. For ecological arguments in favor of novel ecosystems in restoration activities, see Richard J. Hobbs, Eric S. Higgs, and Carol Hall, *Novel Ecosystems: Intervening in the New Ecological World Order* (Oxford: Wiley-Blackwell, 2013). For the opposing view, see Andre F. Clewell and James Aronson, *Ecological Restoration: Principles, Values, and Structure of an Emerging Profession*, 2nd ed. (Washington, DC: Island Press, 2013), 246.

11. William Throop, "The Rationale for Environmental Restoration," in *The Ecological Community: Environmental Challenges for Philosophy, Politics, and Morality*, ed. Roger S. Gottleib (New York: Routledge, 1997); and Luis Balaguer et al., "The Historical Reference in Restoration Ecology: Re-defining a Cornerstone Concept," *Biological Conservation* 176 (2014). Nancy Langston has argued for embracing managed cultural landscapes in restoration activities, rather than attempting to return landscapes to an imagined "original" condition in "Restoration in the American National Forests: Ecological Processes and Cultural Landscapes," in *The Conservation of Cultural Landscapes*, ed. Mauro Agnoletti (Wallingford, UK: CAB International, 2006).

12. *Oxford English Dictionary Online*, s.v. "recovery (*n.*)," definition 4, accessed December 1, 2018, http://www.oed.com/view/Entry/159940.

13. There has been much critique of modern scientific practices by environmental humanities thinkers such as Carolyn Merchant and Val Plumwood, as well as analysis of the role of nature in the Enlightenment, particularly in the work of Lorraine Daston. Emotions, however, have not featured prominently in these analyses.

14. Ursula Heise, *Imagining Extinction: The Cultural Meanings of Endangered Species* (Chicago: University of Chicago Press, 2016), 48.

15. The term first appears in Paul J. Crutzen and Eugene F. Stoermer, "The 'Anthropocene,'" *Global Change Newsletter* 41 (2000): 18–19. It was popularized by Crutzen in "Geology of Mankind," *Nature* 415 (January 3, 2012): 23, and has been a source of debate since then. When the Anthropocene as an era began has been hotly debated, with substantial arguments made for starting at early domestication, industrialization, or atomic detonation. For a review of the potential dates, see Bruce D. Smith and Melinda A. Zeder, "The Onset of the Anthropocene," *Anthropocene* 4 (2013): 8–13, although they come down on the side of early agriculture. For the industrialization argument, see Will Steffen, Paul J. Crutzen, and John R. McNeill, "The Anthropocene: Are Humans Now Overwhelming the Great Forces of Nature?," *Ambio* 36, no. 8 (2007): 614–621.

16. Marsh, *Man and Nature*, iii.

17. Jon Mooallem, *Wild Ones: A Sometimes Dismaying, Weirdly Reassuring Story about Looking at People Looking at Animals in America* (New York: Penguin, 2013), 22.

18. William Stafford, "'This Once Happy Country': Nostalgia for Premodern Society," in *The Imagined Past: History and Nostalgia*, ed. Christopher Shaw and Malcolm Chase (Manchester: Manchester University Press, 1989).

19. Paul Street, "Painting Deepest England: The Late Landscapes of John Linnell and the Uses of Nostalgia," in *The Imagined Past: History and Nostalgia*, ed. Christopher Shaw and Malcolm Chase (Manchester: Manchester University Press, 1989).

20. Lewis Mumford, *Technics and Civilization* (Chicago: University of Chicago Press, 2010), 70.

21. David Lowenthal, "Nostalgia Tells It like It Wasn't," in *The Imagined Past: History and Nostalgia*, ed. Christopher Shaw and Malcolm Chase (Manchester: Manchester University Press, 1989), 21.

22. Jennifer Ladino, *Reclaiming Nostalgia: Longing for Nature in American Literature* (Charlottesville: University of Virginia Press, 2012).

23. Malcolm Chase and Christopher Shaw, "The Dimensions of Nostalgia," in *The Imagined Past: History and Nostalgia*, ed. Christopher Shaw and Malcolm Chase (Manchester: Manchester University Press, 1989), 4.

24. David Lowenthal, "Natural and Cultural Heritage," *International Journal of Heritage Studies* 11, no. 1 (2005).

25. The UNESCO World Heritage program includes "natural" sites as a category. A continuously updated list of sites is available at http://whc .unesco.org/.

26. Peter Coates, "Creatures Enshrined: Wild Animals as Bearers of Heritage," *Past and Present* 226, no. S10 (2015).

27. The Oxford English Dictionary defines "to belong" this way in usage 4(b). See *Oxford English Dictionary Online*, "belong (v.)," accessed December 1, 2018, http://www.oed.com/view/Entry/17506.

28. Emily O'Gorman, "Belonging," *Environmental Humanities* 5 (2014).

29. O'Gorman, "Belonging."

30. Matthew Chew and Andrew Hamilton, "The Rise and Fall of Biotic Nativeness: A Historical Perspective," in *Fifty Years of Invasion Ecology: The Legacy of Charles Elton*, ed. David M. Richardson (Oxford: Blackwell Publishing, 2011), 41; italics in original.

31. See Ian D. Rotherham and Robert A. Lambert, eds., *Invasive and Introduced Plants and Animals: Human Perceptions Attitudes and Approaches to Management* (London: Earthscan, 2011), for some good essays on this topic. In that volume, Chris Smout's "How the Concept of Alien Species Emerged and Developed in 20th-Century Britain" gives an overview of how species have been labeled as not belonging, and Matthew Chew's contribution, "Anekeitaxonomy: Botany, Place and Belonging," traces the development of terms for various states of belonging. The gray squirrel in the United Kingdom is a good example of not belonging, as discussed in Peter Coates's contribution, "Over Here: American Animals in Britain," and in Charles Warren's chapter, "Nativeness and Nation-hood: What Species 'Belong' in Post-devolution Scotland?" Martin Goulding's chapter, "Native or Alien? The Case of the Wild Boar in Britain," reveals the complexities that can arise in labeling with reintro-duced species.

32. For an extended analysis of the junction of species and race in discourse about extinction, see Miles Powell, *Vanishing America: Species Extinction, Racial Peril, and the Origins of Conservation* (Cambridge, MA: Harvard University Press, 2016).

33. David Trigger et al., "Ecological Restoration, Cultural Preferences and the Negotiation of 'Nativeness' in Australia," *Geoforum* 39 (2008).

34. See the essays in part V of Jodi Frawley and Iain McCalman, eds., *Rethinking Invasion Ecologies from the Environmental Humanities* (Abing-don: Routledge, 2014).

35. Thom van Dooren, "Invasive Species in Penguin Worlds: An Ethical Taxonomy of Killing for Conservation," *Conservation and Society* 9, no. 4 (2011): 294.

36. Etienne Benson, "The Urbanization of the Eastern Gray Squirrel in the United States," *The Journal of American History* 100, no. 3 (2013).

37. For histories of the recognition of extinction, see Mark V. Barrow Jr., *Nature's Ghosts: Confronting Extinction from the Age of Jefferson to the Age of Ecology* (Chicago: University of Chicago Press, 2009); and Ryan Tucker Jones, *Empire of Extinction: Russians and the North Pacific's Beasts of the Sea, 1741–1867* (Oxford: Oxford University Press, 2014).

38. I recognize that in making this list, I am actively mobilizing both the elegiac narrative and epic cataloging impulse that Heise discusses so well in chapter 2 of *Imagining Extinction*. I agree with her conclusion in that chapter that the meaning of the data on the list "emerges only through the stories we tell about them" (86); thus, all these names really need stories to be meaningful.

39. The story of how the Japanese wolf went from being deified to purposefully hunted to extinction is told in Brett L. Walker, *The Lost Wolves of Japan* (Seattle: University of Washington Press, 2005).

40. Hayden Fowler, "Epilogue: New World Order—Nature in the Anthropocene," in *Animals in the Anthropocene: Critical Perspectives on Non-Human Futures*, ed. Human Animal Research Network Editorial Collective (Sydney: Sydney University Press, 2015), 245. On potential forward-looking responses to loss in the Anthropocene, see Katherine Gibson, Deborah Bird Rose, and Ruth Fincher, "Preface," in *Manifesto for Living in the Anthropocene*, ed. Katherine Gibson, Deborah Bird Rose, and Ruth Fincher (New York: Punctum Books, 2015).

41. Glenn Albrecht et al., "Solastalgia: The Distress Caused by Environmental Change," *Australasian Psychiatry* 15, no. S1 (2007).

42. Lesley Head, *Hope and Grief in the Anthropocene: Re-conceptualising Human-Nature Relations* (Abingdon: Routledge, 2016), 6; italics in original.

43. Deborah Bird Rose, Thom van Dooren, and Matthew Chrulew, eds., *Extinction Studies: Stories of Time, Death, and Generations* (New York: Columbia University Press, 2017).

44. Thom van Dooren, *Flight Ways: Life and Loss at the Edge of Extinction* (New York: Columbia University Press, 2014).

45. Sarah M. Pike, "Mourning Nature: The Work of Grief in Radical Environmentalism," *Journal for the Study of Religion, Nature and Culture* 10, no. 4 (2016) 419–441.

46. Ashlee Cunsolo and Neville R. Ellis, "Ecological Grief as a Mental Health Response to Climate Change-Related Loss," *Nature Climate Change* 8 (2018); and the essays in Ashlee Cunsolo and Karen Landman, eds., *Mourning Nature: Hope at the Heart of Ecological Loss and Grief* (Montreal: McGill-Queen's University Press, 2017). See also Lesley Head, "The Anthropoceneans," *Geographical Research* 53, no. 3 (2015), for a discussion of the need to acknowledge our grief as we face climate change. Todd Walton and Wendy S. Shaw, "Living with the Anthropocene Blues," *Geoforum* 60 (2015) makes a good case for investigating the variety of psychological responses to the stress of environmental change.

47. Richard J. Hobbs, "Grieving for the Past and Hoping for the Future: Balancing Polarizing Perspectives in Conservation and Restoration," *Restoration Ecology* 21, no. 2 (2013): 146.

48. Elizabeth Kolbert, "How to Write about a Vanishing World," *New Yorker*, October 15, 2018.

49. Laura Smith, "On the 'Emotionality' of Environmental Restoration: Narratives of Guilt, Restitution, Redemption and Hope," *Ethics, Policy & Environment* 17 (2014): 304.

50. For work on emotional communities, see Barbara Rosenwein, *Emotional Communities in the Early Middle Ages* (Ithaca, NY: Cornell University Press, 2006); and Barbara Rosenwein, "Problems and Methods in the History of Emotions," *Passions in Context* 1 (2010). Peter N. Stearns and Carol Z. Stearns, in "Emotionology: Clarifying the History of Emotions and Emotional Standards," *American Historical Review* 90, no. 4 (1985), coin *emotionology* as the study of emotional standards. For examinations of one particular emotion, see, for example, Barbara H. Rosenwein, ed., *Anger's Past: The Social Uses of an Emotion in the Middle Ages* (Ithaca, NY: Cornell University Press, 1989); David Konstan, *Pity*

Transformed (London: Duckworth, 2001); and Joanna Bourke, *Fear: A Cultural History* (London: Virago, 2005).

51. Quoted in Frank Biess, "Forum: History of Emotions," *German History* 28, no. 1 (2010): 70. For an example of a discourse analysis approach to extinction, see Amy Lynn Fletcher, *Mendel's Ark: Biotechnology and the Future of Extinction* (Dordrecht: Springer Science, 2014), which looks at discourses of extinction in relationship to biotechnology-driven de-extinction science.

52. Monique Scheer, "Are Emotions a Kind of Practice (And Is That What Makes Them Have a History)? A Bourdieuian Approach to Understanding Emotion," *History and Theory* 51 (2012): 195.

53. Quoted in Biess, "Forum: History of Emotions," 71.

54. For a good overview of the source possibilities, see Susan Broomhall, ed., *Early Modern Emotions: An Introduction* (London: Routledge, 2017).

55. Andrea Gaynor introduced her project in "Environmental History and the History of Emotions," Histories of Emotion research group, https://historiesofemotion.com/2017/06/16/environmental-history-and-the-history-of-emotions/, and has been working on an environmental history article to publish on emotional reactions to urban frogs in Australia. Michael Egan has introduced his approach to toxic fear in "Chemical Unknowns: Preliminary Outline for an Environmental History of Fear," in *Framing the Environmental Humanities*, ed. Hannes Bergthaller and Peter Mortensen (Leiden: Brill, 2018).

56. Daniel Macfarlane, "Emotional and Environmental History at Niagara Falls," Otter blog, Network in Canadian History & Environment, http://niche-canada.org/2017/09/28/emotional-and-environmental-history-at-niagara-falls/.

57. Yi-Fu Tuan, *Landscapes of Fear* (New York: Pantheon Book, 1979).

58. Karl Jacoby's *Crimes against Nature: Squatters, Poachers, Thieves, and the Hidden History of American Conservation* (Berkeley: University of California Press, 2001) offers insightful analysis of the moral ecology of rural Americans and their relationship to wildlife. Powell's *Vanishing*

America gives a detailed treatment of how racist beliefs structured the moral ecology of elite conservationists.

59. Examples include Nicole Seymour, "Toward an Irreverent Ecocriticism," *Journal of Ecocriticism* 4, no. 2 (2012); Adrian J. Ivakhiv, *Ecologies of the Moving Image: Cinema, Affect, Nature* (Waterloo, Canada: Wildred Laurier University Press, 2013); Alexa Weik von Mossner, ed., *Moving Environments: Affect, Emotion, Ecology, and Film* (Waterloo, Ontario: Wilfrid Laurier University Press, 2014); Alexa Weik von Mossner, *Affective Ecologies: Empathy, Emotion, and Environmental Narrative* (Columbus: Ohio State University Press, 2017); and Kyle Bladow and Jennifer Ladino, eds., *Affective Ecocriticism: Emotion, Embodiment, Environment* (Lincoln: University of Nebraska Press, 2018). I will note that often the word *affect* is used in these academic works for not only the internal, physical response but also the outward-facing response, which tends to blur the lines between affect and emotion.

60. For studies of wildlife conservation actions to avoid extinction all together, see Thomas R. Dunlap, *Saving America's Wildlife* (Princeton, NJ: Princeton University Press, 1988), for a foundational reading of the development of wildlife policy in the United States; Tina Loo, *States of Nature: Conserving Canada's Wildlife in the Twentieth Century* (Vancouver: UBC Press, 2006), for the complexities of wildlife conservation and how both governments and private individuals are involved; Peter S. Alagona, *After the Grizzly: Endangered Species and the Politics of Place in California* (Berkeley: University of California Press, 2013), for a focused discussion of habitat conservation efforts; Jamie Lorimer, *Wildlife in the Anthropocene: Conservation after Nature* (Minneapolis: Minnesota University Press, 2015), for the biopolitics of conservation; and E. Elena Songster, *Panda Nation: The Construction and Conservation of China's Modern Icon* (Oxford: Oxford University Press, 2018), for an example of how an endangered species can become iconic.

61. Philip J. Seddon, "From Reintroduction to Assisted Colonization: Moving Along the Conservation Translocation Spectrum," *Reintroduction Ecology* 18, no. 6 (2010); and Dolly Jørgensen, "What's History Got to Do with It? A Response to Seddon's Definition of Reintroduction," *Restoration Ecology* 19 (2011).

62. Barrow, *Nature's Ghosts*, chap. 4.

63. Hall, "Restoration and the Search for Counter-narratives," 317.

64. On redemption as a key component of restoration activities, see Smith, "On the 'Emotionality' of Environmental Restoration."

65. See Dolly Jørgensen, "Rethinking Rewilding," *Geoforum* 65 (2015), for a tracing of the concept of rewilding in scientific literature.

66. See, for example, Frans Schepers and Paul Jepson, "Rewilding in a European Context," *International Journal of Wilderness* 22, no. 2 (2016), for the way in which rewilding advocates distinguish themselves as part of a conservation movement. Schepers is a cofounder of Rewilding Europe.

67. For analysis of the Oostvaardersplassen experiment, see Lorimer, *Wildlife in the Anthropocene*; and Jamie Lorimer and Clemens Driessen, "Wild Experiments at the Oostvaardersplassen: Rethinking Environmentalism in the Anthropocene," *Transactions of the Institute of Royal Geographers* 39, no. 2 (2014). On Pleistocene Park's rewilding initiative, which is also tied to the potential resurrection of the mammoth itself, see Matthew Chrulew, "Reversing Extinction: Restoration and Resurrection in the Pleistocene Rewilding Projects," *Humanimalia* 2, no. 2 (2011).

68. See Pike's discussion of "the redemptive promises of 'the wild'" to counter the mourning of environmental loss and feelings of guilt by environmentalists in "Mourning Nature," 436–438.

69. Three articles by Jamie Lorimer and Clemens Driessen discuss the bovine politics of using Heck cattle, a breed created under the German Nazi regime, in rewilding efforts: Lorimer and Driessen, "Bovine Biopolitics and the Promise of Monsters in the Rewilding of Heck Cattle," *Geoforum* 48 (2013); Driessen and Lorimer, "Back-Breeding the Aurochs: The Heck Brothers, National Socialism and Imagined Geographies for Nonhuman Lebensraum," in *Hitler's Geographies*, ed. P. Giaccaria and C. Minca (Chicago: University of Chicago Press, 2016); and Lorimer and Driessen, "From 'Nazi Cows' to Cosmopolitan 'Ecological Engineers': Specifying Rewilding through a History of Heck Cattle," *Annals of the*

American Association of Geographers 106, no. 3 (2016). For the quagga project, see Sandra Swart, "Zombie Zoology: History and Reanimating Extinct Animals," in *The Historical Animal*, ed. Susan Nance (Syracuse, NY: Syracuse University Press, 2015).

70. For details about the scientific process of resurrecting species with genetic techniques, see Beth Shapiro, *How to Clone a Mammoth: The Science of De-extinction* (Princeton, NJ: Princeton University Press, 2015).

71. The linkage between de-extinction and the science fiction work *Jurassic Park* is not entirely misplaced. Heise notes that de-extinction as a practice shifts the narrative from the elegiac mode to the science fiction mode, potentially opening new ways of envisioning future ecosystems that are similar to science fiction narratives of terraforming and multispecies worlds; see chapter 6 of *Imagining Extinction*.

72. Steven Spielberg, dir., *Jurassic Park* (Universal City, CA: Universal Pictures, 1993).

73. Carl Zimmer, "Bringing Them Back to Life," *National Geographic* (April 2013), https://www.nationalgeographic.com/magazine/2013/04/species-revival-bringing-back-extinct-animals/.

74. The result of the cloning experiment was not reported in a scientific journal until 2009: J. Folch et al., "First Birth of an Animal from an Extinct Subspecies (*Capra pyrenaica pyrenaica*) by Cloning," *Theriogenology* 71 (2009).

75. See, for example, Ben A. Minteer, "When Extinction Is a Virtue," in *After Preservation: Saving American Nature in the Age of Humans*, ed. Ben A. Minteer and Stephen J. Pyne (Chicago: University of Chicago Press, 2015); and Ronald Sandler's works "The Ethics of Reviving Long Extinct Species," *Conservation Biology* 28, no. 2 (2013), and "Techno-Conservation in the Anthropocene: What Does It Mean to Save a Species?," in *The Routledge Companion to the Environmental Humanities*, ed. Ursula Heise, Jon Christensen, and Michelle Niemann (Abingdon: Routledge, 2017).

76. See Fletcher, *Mendel's Ark*, for a discussion of technoscience as a response to extinction. I have previously argued that human interaction

with animals has long been based on technology: Dolly Jørgensen, "Not by Human Hands: Five Technological Tenets for Environmental History in the Anthropocene," *Environment and History* 20 (2014).

77. I am certainly not the first scholar to write about the extinction history of the passenger pigeon, but rather than focusing on the extinction event itself, I am interested here in analyzing the emotional framework of grief evident in writers who reflected upon the extinction, which is a new approach. For a significant treatment of the inherent ties between capitalism and the extinction of the passenger pigeon, see chapter 1 of Jennifer Price, *Flight Maps: Adventures with Nature in Modern America* (New York: Basic Books, 1999). Recent monographs on the passenger pigeon extinction include Mark Avery, *A Message from Martha: The Extinction of the Passenger Pigeon and Its Relevance Today* (New York: Bloomsbury, 2014); Errol Fuller, *The Passenger Pigeon* (Princeton, NJ: Princeton University Press, 2015); and Joel Greenberg, *A Feathered River across the Sky: The Passenger Pigeon's Flight to Extinction* (New York: Bloomsbury USA, 2014). See also the discussion of the passenger pigeon's extinction that is interwoven with the narrative of the near extinction of the American bison in Barrow, *Nature's Ghosts*.

78. Carrie Friese, "Cloning in the Zoo: When Zoos Become Parents," in *The Ark and Beyond: The Evolution of Zoo and Aquarium Conservation*, ed. Ben A. Minteer, Jane Maienschein, and James P. Collins (Chicago: University of Chicago Press, 2018), 274. See also Stephanie S. Turner, "Open-ended Stories: Extinction Narratives in Genome Time," *Literature and Medicine* 26, no. 1 (2007), for a discussion of the way that de-extinction modifies the time-based narrative of extinction by positing that it can be reversed.

79. Jan Assmann, *Moses the Egyptian: The Memory of Egypt in Western Monotheism* (Cambridge, MA: Harvard University Press, 1997), 9.

80. The "cultural turn" in environmental history—a focus on discourse, story, and narrative in environmental history—emerged around the turn of the millennium and is still going strong. See Richard White, "From Wilderness to Hybrid Landscapes: The Cultural Turn in Environmental History," *Historian* 66, no. 3 (2004), for early reflections on this development.

2 Reintroducing

1. The release story is told in Eric Festin, "Bäverns återinplantering," *Jämten* 16 (1922); and Sven Arbman, "När bäfvern återinfördes i Bjurälfen," *Svenska Jägareförbundets Tidskrift* 60 (1922). All translations from Norwegian and Swedish in this book are my own.

2. The Marquis of Bute had released Canadian beavers into an enclosure on the Isle of Bute, a Scottish island west of Glasgow in 1874. They died by the 1880s. Contemporary writers referred to the project as a "reintroduction," including James Edmund Harting in *British Animals Extinct within Historic Times* (Boston: J. R. Osgood, 1880), but it was not technically a European beaver reintroduction because the marquis introduced a different species of beaver and they were never living outside of the enclosure.

3. Jonathan L. Freedman, Sue A. Wallington, and Evelyn Bless, "Compliance without Pressure: The Effect of Guilt," *Journal of Personality and Social Psychology* 7 (1967): 117.

4. Melissa S. Burnett and Dale A. Lunsford, "Conceptualizing Guilt in the Consumer Decision-Making Process," *Journal of Consumer Marketing* 11, no. 3 (1994); and Ann P. Minton and Randall L. Rose, "The Effects of Environmental Concern on Environmentally Friendly Consumer Behavior: An Exploratory Study," *Journal of Business Research* 40, no. 1 (1997).

5. Mark A. Ferguson and Nyla R. Branscombe, "Collective Guilt Mediates the Effect of Beliefs about Global Warming on Willingness to Engage in Mitigation Behavior," *Journal of Environmental Psychology* 30 (2010).

6. Most people think there is one kind of beaver globally, but the beaver of Eurasia (*Castor fiber*) and the beaver of North America (*Castor canadensis*) are different species. In fact, they are *very* different species. They have a different number of chromosomes (forty-eight in the European and forty in the North American). There is no known hybridization between the two species, and it is assumed that they cannot produce viable offspring. They diverged genetically about 7.5 million

years ago after the early beaver migrated from Asia to North America. See Susanne Horn et al., "Mitochondrial Genomes Reveal Slow Rates of Molecular Evolution and the Timing of Speciation in Beavers (Castor), One of the Largest Rodent Species," *PLOS One* 6, no. 1 (2011).

7. Modern beaver reintroduction projects typically do not target beaver conservation, but rather are interested in promoting the ecosystem modifications that beavers cause, particularly the increase of wetted habitats at the liminal zone between water and land. However, the scale of ecological impact of beavers may vary based on the geographical location: see Frank Rosell et al., "Ecological Impact of Beavers *Castor fiber* and *Castor canadensis* and Their Ability to Modify Ecosystems," *Mammal Review* 35, no. 3–4 (2005).

8. For a good introduction to the historical relations between humans and beavers, both in Europe and North America, see Rachel Poliquin, *Beaver* (London: Reaktion Books, 2015).

9. Gerald of Wales, *The Itinerary through Wales and the Description of Wales*, ed. W. Llewelyn Williams (London: J. M. Dent, 1912), chapter 3.

10. For the medieval Russian fur trade, see Thomas S. Noonan and Roman K. Kovalev, "'The Furry 40s': Packaging Pelts in Medieval Northern Europe," in *States, Societies, Cultures. East and West: Essays in Honor of Jaroslaw Pelenski*, ed. Janusz Duzinkiewicz (New York: Ross Publishing, 2004). Archeological evidence from Birka, Sweden, shows that processing and trade of fur, including beaver fur, was carried out there before the site was abandoned in the tenth century: see Bengt Wigh, "Animal Bones from the Viking Town of Birka, Sweden," in *Leather and Fur: Aspects of Early Medieval Trade and Technology*, ed. Esther Cameron (London: Archetype Publications, 1998).

11. Nils Gisler, "Rön och berättelse om Bäfverns natur, hushållning och fångande," *Kungl. Svenska vetenskapsakademiens handlingar* 17 (1756): 220.

12. John Russell, *Boke of Nurture*, ed. Frederick J. Furnivall (Bungay, UK: John Childs and Son, 1867), line 547.

13. Gerald of Wales, *Itinerary through Wales*, chap. 3.

14. S. Nilsson, *Skandinavisk Fauna*, vol. 1, *Däggdjuren* (Lund: Gleerups, 1847), 419.

15. Pliny the Elder, *The Natural History*, 6 vols., trans. John Bostock and H. T. Riley (London: Henry G. Bohn, 1855–1857), book 32, chap. 13.

16. Jan Karlsen, "Moskus, Zibet, Castoreum og Ambra: Animalske droger i det gamle apotek," *Cygnus—en norsk farmahistorisk skriftserie* 10 (October 2004).

17. Johan Winter, "Et och annat om Bäfvern," *Svenska Jägarförbundets Nya Tidskrift* 11 (1873). A fuller version of the account was printed later as part of Erik Modin, "Anteckningar om bäfvern, dess förekomst och fångst m. m. i Västerbotten under förra hälften af 1800-talet," *Svenska Jägarförbundets Nya Tidskrift* 45 (1907). Winter's original piece concluded by saying that beaver were still in Lappland, but Modin's version replaced that with a paragraph stating that overhunting with nets and conversion of land into agricultural use caused the beaver to become extinct.

18. Alarik Behm, *Nordiska Däggdjur: 177 bilder från Skansen* (Uppsala: J. A. Lindblads Bokförlags Aktiebolag, 1922), 131.

19. Gisler, "Rön och berättelse om Bäfverns natur, hushållning och fångande," 221.

20. G. Swederus, *Skandinaviens Jagt: Djurfänge och Vildafugl* (Stockholm: P. A. Norstedt & Söner, 1832), 133.

21. Nilsson, *Skandinavisk Fauna*, vol. 1, 416.

22. F. Unander, "Ett från svenska jagtbanan försvunnet dyrbart djur," *Svenska Jägarförbundets Nya Tidskrift* 11 (1873): 33.

23. Duncan Halley and Frank Rosell, "Population and Distribution of European Beavers (*Castor fiber*)," *Lutra* 46, no. 3 (2003): 91–94.

24. He uses the Swedish words *införande* (literally, "introduce") and *inplantering* ("implantation") when talking about the beaver, which are the most common everyday words used in Swedish when discussing what scientists would label as *reintroduction*. Although I use *reintroduce* in this chapter, the historical actors would have chosen other words to

represent the same action. Modin's suggestion was to bring in Canadian beavers, which are not the same beaver species, although they likely would have filled the same ecologic and economic niche as European beavers.

25. Erik Modin, "Bör ej nägot göras för bäfverns återinförade i vårt land?" *Svenska Jägareförbundets Tidskrift* 49 (1911).

26. Modin, "Bör ej nägot göras för bäfverns återinförade i vårt land?" 193.

27. Modin, "Bör ej nägot göras för bäfverns återinförade i vårt land?" 194.

28. Eric Festin, "Jubileumsutställningen och Kulturmässan i Östersund 1920," *Jämten* 14 (1920).

29. His speech was printed afterward as Alarik Behm, "Några ord om naturskydd," *Jämten* 14 (1920), for the local audience; and as Alarik Behm, "Naturskydd, särskilt i Jämtland," *Sveriges Natur* (1921), for the national audience.

30. Carin Taflin, "Mina år med Eric Festin," *Jämten* 79 (1986).

31. Karl-Erik Forsslund, *Hembygdsvård*, 2 vols. (Stockholm: Wahlström & Widstrand, 1914).

32. Forsslund, *Hembygdsvård*, 1:35.

33. Behm, "Några ord om naturskydd," 42. These sentiments are mirrored in German environmental protection, which also stresses the landscape (*Landschaft*) and its natural and cultural components, as discussed in Heise, *Imagining Extinction*, 103–107.

34. Eric Festin, "Bäverns återinplantering i Jämtland," *Sveriges Natur* 12 (1921).

35. Modin, "Anteckningar om bäfvern."

36. Eric Festin, "Fridlysning av Bjurälvdalens karstlandskap och återinplantering av bävern: En samtidig lösning av två viktiga naturskyddsfrågor," *Sveriges Natur* 13 (1922): 60.

37. Festin, "Fridlysning av Bjurälvdalens karstlandskap och återinplantering av bävern," 57.

38. Festin, "Fridlysning av Bjurälvdalens karstlandskap och återinplantering av bävern," 59.

39. Festin, "Bäverns återinplantering," 84.

40. Festin, "Fridlysning av Bjurälvdalens karstlandskap och återinplantering av bävern," 59.

41. Festin, "Bäverns återinplantering i Jämtland," 148.

42. Fredr Svenonius, "Bjurälfdalens karstlandskap i norra Jämtland," *Sveriges Natur* 1 (1910). Beaver structures and water channel modifications can last for centuries. See Bryony Coles, *Beavers in Britian's Past*, for a discussion of beaver archeology.

43. Festin, "Bäverns återinplantering i Jämtland," 148.

44. Festin, "Bäverns återinplantering i Jämtland," 148; italics in original.

45. Robert Collett, "Meddelelser om Norges Pattedyr i Aarene 1876–1881," *Nyt Magazin for Naturvidenskaberne* 27 (1882); and Robert Collett, "Om Bæveren (Castor fiber), og dens Udbredelse i Norge fordum og nu," *Nyt Magazin for Naturvidenskaberne* 28 (1883).

46. Collett, "Om Bæveren (Castor fiber)," 44; and Robert Collet, "Bæveren i Norge, dens Udbredelse og Levemaade (1896)," in *Bergens museums aarbog* (1897): article 1, 85.

47. The beaver population along the Nid had survived and expanded largely due to Nicolai Aall, owner of the Næs Ironworks in Tvedestrand. Some time after he took over the ironworks in 1844 upon the death of his father Jacob Aall, Nicolai banned hunting and trapping of beavers on property owned by the company. The ban applied not only to the area immediately surrounding the mill but also the extensive landholdings of the company forty kilometers inland around Åmli, which were used to supply trees for charcoal needed in the ironworking process. We do not know the exact reason Nicolai decided to implement the ban, but it was likely tied to his extensive personal interest in hunting.

Like American President Theodore Roosevelt, who was both a big game hunter and a leader in nature protection, Nicolai was considered a "friend of wildlife" at a time when killing and protecting were understood as two sides of the same coin. For a short biography, see Haagen Krog Steffens, "Nicolai Benjamin Aall," in *Slægten Aall* (Centraltrykkeriet: Kristiania, 1908).

48. Sigvald Salvesen, "The Beaver in Norway," *Journal of Mammalogy* 9 (1928).

49. Letter from Sven Arbman to Eric Festin, dated June 21, 1921, folder C32.2 Bäverinplanteringen, subfolder 1920–1922, Jamtli archive.

50. Sometimes he used Jensen-Tveit, and at other times wrote his last name Jenssen or Jenssen-Tveit. I will use Jensen throughout for consistency.

51. Letter from Peder M. Jensen to Eric Festin, dated September 20, 1921, folder C32.2 Bäverinplanteringen, subfolder 1920–1922, Jamtli archive.

52. Letter from Alarik Behm to Eric Festin, no date (likely August/September 1921), folder C32.2 Bäverinplanteringen, subfolder 1920–1922, Jamtli archive.

53. Arne Tjomsland, Dødsfall, *Aftenposten*, October 1, 1963, 11.

54. "Bäverkärlek," *Dagens Nyheter*, October 24, 1935, newspaper clipping in folder E.5. Natur- och Djurskydd 1935–1937, Jamtli archive.

55. Letter from Gustav Werner to Eric Festin, dated May 24, 1921, folder C32.2 Bäverinplanteringen, subfolder 1920–1922, Jamtli archive.

56. The folder C32.2 Bäverinplanteringen, subfolder 1922, in the Jamtli archive is filled with "Bidrag till bäverfonden" sign-up sheets.

57. Arbman, "När bäfvern återinfördes i Bjurälfen," 275.

58. Festin, "Bäverns återinplantering," 85. Skansen would do the same for a later beaver reintroduction project in the county of Västerbotten.

59. Festin, "Bäverns återinplantering," 90.

60. Festin, "Bäverns återinplantering," 86.

61. Festin, "Bäverns återinplantering," 85–86.

62. The photos are in the Jamtli photo archive. The series by Thomasson includes NTh621 to NTh626 and NTh7169. I have written about the box as a domestication tool in Dolly Jørgensen, "Muskox in a Box and Other Tales of Containers as Domesticating Mediators in Animal Relocation," in *Animal Housing and Human-Animals Relations: Politics, Practices and Infrastructures*, ed. Tone Druglitrø and Kristian Bjørkdahl (Abingdon: Routledge, 2016).

63. Festin, "Bäverns återinplantering," 90.

64. In the Jamtli photo archive, the images by Festin are C1405 to C1407; those by Nils Thomasson are NTh23047 to NTh23058, NTh23209, NTh23210, and NTh2312.

65. Letter from Aktibolaget Svensk Filmindustri to Eric Festin, dated June 27, 1922, folder C32.2 Bäverinplanteringen, subfolder 1920–1922, Jamtli archive.

66. In 1925, Sigvald Salvesen successfully had a "school film" about beavers produced in Åmli. In a letter requesting permission to capture beavers in order to get close up shots, Salvesen claimed that he had "over many years thought about making such a film" but encountered many difficulties. See letter from Salvesen to Landbruksdepartmentet, dated September 18, 1925, Archive RA/S-6087/D/Da/Dab/L0089, National Archives of Norway. A review of the film in the *Stavanger Aftenblad* on December 16, 1925, praised it as an educational film, encouraging parents to take their children to see "Norway's first beaver film."

67. Eric Festin's handwritten protocol from the release, folder C32.2 Bäverinplanteringen, subfolder 1920–1922, Jamtli archive.

68. Arbman, "När bäfvern återinfördes i Bjurälfen," 278.

69. Forsslund, *Hembygdsvård*, 2:109.

70. Festin, "Fridlysning av Bjurälvdalens karstlandskap och återinplantering av bävern," 62; italics in original.

71. Eric Festin, "Sveriges nya bäverstam: Det första återinplanterings-initiativet och dess efterföljare," *Jämtlands läns jaktvårdsförening årsbok* (1928).

72. Eric Festin, "Sveriges nya bäverstam," *Sveriges Natur* 19 (1928).

73. A. Sylvén, Gunnar Beronius, and Nils Almlöf, "Till våra lärare!," *Västerbottens läns jaktvårdsförening årsbok* (1921).

74. Nils G. Ringstrand, "Jaktvård," *Västerbottens läns jaktvårdsförening årsbok* (1921).

75. Loo, *States of Nature*, 26–27.

76. Thomas R. Dunlap, "Sport Hunting and Conservation, 1880–1920," *Environmental History Review* 12 (1988): 57.

77. European hunting in the eighteenth century likewise was focused on domination of nature as an instrument to communicate power; see Martin Knoll, "Hunting in the Eighteenth Century: An Environmental History Perspective," *Historical Social Research* 29 (2004).

78. Dunlap, "Sport Hunting and Conservation," 54.

79. A. Sylvén, "Bävern tillbaka till Västerbotten," *Västerbottens läns jaktvårdsförening årsbok* (1922).

80. A. Sylvén, "Bävern tillbaka till Västerbotten," 39.

81. A. Sylvén, "Våra bävrar," *Västerbottens läns jaktvårdsförening årsbok* (1924): 7.

82. A. Sylvén and Lennart Wahlberg, "Årsmötet 1924," *Västerbottens läns jaktvårdsförening årsbok* (1924).

83. "Ytterligare fyra bävrar nu inplanterade i Tärna-ån," *Västerbottens-Kuriren*, July 25, 1924.

84. Axel Anderson, "Då bävern återbördades till Västerbotten," *Västerbotten: Västerbottens läns hembygdsförenings årsbok* 5 (1924–1925): 284.

85. Lennart Wahlberg, "Bäverns återbördande till Syd-Lappland," *Västerbottens läns jaktvårdsförening årsbok* (1925). A later article refers to

the beavers as the "wards" of Lennart Wahlberg, who was listed as the "officiant" in the release of the four beavers on July 20, 1924: G. H. von Post, "På besök hos Västerbottens bävrar," *Västerbottens läns jaktvårds-förening årsbok* (1930).

86. Post, "På besök hos Västerbottens bävrar," 64.

87. For example, *Västerbottens läns jaktvårdsförening årsbok* (1930): 54.

88. S. Deinet et al., *Wildlife Comeback in Europe: The Recovery of Selected Mammal and Bird Species*, final report to Rewilding Europe by ZSL, BirdLife International, and the European Bird Census Council (London: ZSL, 2013).

89. Carl Fries, *Bäverland: En bok om bävern och hans verk* (Stockholm: Nordisk Rotogravyr, 1940), 140.

90. See, for example, Ingvar Ling, "Bävern i Norrbotten," *Norrbottens Natur* 37 (1981).

91. Göran Hartman, "The Beaver (*Castor fiber*) in Sweden," in *Restoring the European Beaver: 50 Years of Experience*, ed. Göran Sjöberg and John P. Ball (Sofia-Moscow: Pensoft, 2011).

92. Letter from Skogdirektør to Gustaf Kolmodin, dated October 8, 1926, archive RA/S-6087/D/Da/Dab/L0089, National Archives of Norway.

93. This appears to be quite different from the regular complaints in the 1940s leveled against beavers that had been reintroduced in California in the late 1930s, discussed in Steven M. Fountain, "Ranchers' Friend and Farmers' Foe: Reshaping Nature with Beaver Reintroduction in California," *Environmental History* 19 (2014).

94. Erik Geete, "Bävern i Sverige och Norge," *Skogen* 16 (1929): 300; italics in original.

95. Dan Flores, *American Serengeti: The Last Big Animals of the Great Plains* (Lawrence: University Press of Kansas, 2016), 7–8; italics in original.

3 Rewilding

1. O. A., "Dovrefjellenes nyeste pryd—Moskus-dyrene trives godt," *Aftenposten*, October 19, 1940.

2. Jørgensen, "Rethinking Rewilding."

3. Tourism entrepreneurs are one of the focuses for development under the Rewilding Europe scheme. See, for example, Rewilding Europe, *2015 Annual Review*, https://www.rewildingeurope.com/wp-content/uploads/publications/rewilding-europe-annual-review-2015/index.html;   it devotes an eleven-page section to enterprise development.

4. C. Josh Donlan et al., "Pleistocene Rewilding: An Optimistic Agenda for Twenty-first Century Conservation," *American Naturalist* 168, no. 5 (November 2006): 661. A shorter conceptual piece was published by the group initially in the major journal *Nature*: C. Josh Donlan et al., "Rewilding North America," *Nature* 436 (2005).

5. Donlan et al., "Pleistocene Rewilding," 673. The Donlan et al. proposal proved hugely controversial, with an outpouring of both scientist and layperson sentiments for and against the idea of rewilding North America. For summaries of how the proposers felt about the reception of their idea, see C. Josh Donlan and Harry W. Greene, "NLIMBY: No Lions in My Backyard," in *Restoration and History: The Search for a Usable Environmental Past*, ed. Marcus Hall (New York: Routledge, 2010); and Harry W. Greene, "Pleistocene Rewilding and the Future of Biodiversity," in *After Preservation: Saving American Nature in the Age of Humans*, ed. Ben A. Minteer and Stephen J. Pyne (Chicago: University of Chicago Press, 2015).

6. Higgs, *Nature by Design*, 143.

7. George Monbiot, *Feral: Searching for Enchantment on the Frontiers of Rewilding* (London: Penguin, 2013). The version of the book issued in the United States has a different title: *Feral: Rewilding the Land, the Sea, and Human Life* (Chicago: University of Chicago Press, 2014).

8. Monbiot, *Feral*, 7.

9. See Claudia Bloeser and Titus Stahl, "Hope," in *The Stanford Encyclopedia of Philosophy* (Spring 2017 edition), ed. Edward N. Zalta, for an overview of philosophical writings on hope.

10. I follow Jayne M. Waterworth's reading here that hope is an anticipation (it might or might not come true) rather than an expectation (which assumes that it will come true); see Waterworth, *A Philosophical Analysis of Hope* (New York: Palgrave Macmillan, 2004), 8–14.

11. See James R. Averill, George Catlin, and Kynm Koo Chon, *Rules of Hope* (New York: Springer Science & Business Media, 1990), chap. 1.

12. Joseph J. Godfrey, *A Philosophy of Human Hope* (Dordrecht: Marinus Nijhoff, 1987), 169.

13. Godfrey, *A Philosophy of Human Hope*, 8.

14. Waterworth, *A Philosophical Analysis of Hope*, 5–6.

15. For the transgressive possibilities of hope, see Patrick Shade, *Habits of Hope: A Pragmatic Theory* (Nashville: Vanderbilt University Press, 2001); and Shade, "Shame, Hope and the Courage to Transgress," in *Theories of Hope: Exploring Alternative Affective Dimensions of Human Experience*, ed. Rochelle Green (London: Rowman & Littlefield, 2019).

16. See Lisa Kretz, "Hope in Environmental Philosophy," *Journal of Agricultural and Environmental Ethics* 26, no. 5 (2013), for a discussion of the philosophical importance of hope in responding to ecological crisis.

17. Smith, "On the 'Emotionality' of Environmental Restoration."

18. See Bourke, *Fear*.

19. Egan, "Chemical Unknowns."

20. Bourke, *Fear*, 354.

21. For general muskox biology and natural history, see Alwin Pedersen, *Der Moschusochs* (Wittenberg Luterstadt: A. Ziemsen, 1958); and Peter Lent, *Muskoxen and Their Hunters: A History* (Norman: University of Oklahoma Press, 1999).

22. Pierre François Xavier de Charlevoix, *Journal d'un voyage fait par ordre du roi dans l'Amerique septentrionnale*, vol. 3 (Paris: Chez Nyon, 1744), 132.

23. M. H. de Blainville, "Sur plusieurs espéces d'animaux mammiferes, de l'ordre des ruminants," *Bulletin des Sciences, par La Société Philomatique de Paris* (1816).

24. Note that although the earlier spelling was *Ovibus* (with a *u*), the currently accepted name for the species is *Ovibos* (with an *o*). Both are used in the historical literature.

25. Vilhjalmur Stefansson, *The Northward Course of Empire* (New York: Harcourt, Brace and Co., 1922), 139.

26. "Ingen statsimport av moskus som husdyr," *Aftenposten*, January 30, 1969, 34; and "Polar-sauen igjen på de nord-norske beiter," *Aftenposten*, November 28, 1970, 1. Otto Sverdrup also apparently called the animal *polarfe* (literally "polar cattle") earlier: "Skal moskusdyrene for alvor holde sitt inntog på vårt høifjell," *Aftenposten*, May 11, 1933, 4.

27. "Rester af en moskusoxe i Norge," *Aftenposten*, May 21, 1913, 4; and P. A. Øyen, "Fund af mammut og moskusoxe i vort land," *Aftenposten*, May 13, 1917, 1.

28. Adolf Hoel, "Moskusoksen: bestand, jakt, fangst, omplantnings-forsøk," unpublished manuscript, SM-5138, Norsk Polar Institute archive.

29. For a more thorough discussion of these motivations, see Peder Roberts and Dolly Jørgensen, "Animals as Instruments of Norwegian Imperial Authority in the Interwar Arctic," *Journal for the History of Environment and Society* 1 (2016); and Dolly Jørgensen, "Migrant Muskox and the Naturalization of National Identity in Scandinavia," in *The Historical Animal*, ed. Susan Nance (Syracuse, NY: Syracuse University Press, 2015).

30. Adolf Hoel, "Overføring av moskusokser til Svalbard," *Norge tidsskrift om vårt land* 6, no. 1 (1930). Although eating muskox was stated as an eventual goal, killing the muskoxen was strictly prohibited when they were released. A sign was set up in the Svalbard mining company's

shop in Longyearbyen, stating: "Muskoxen are protected." Although six muskoxen were shot in the winter of 1942–1943 as food for the military garrison on Svalbard and the herd in the Dovre mountains was killed off during the WWII German occupation of Norway, the muskox remained a protected animal, illegal to hunt.

31. Landsforeningen, "Moskusoksen bør vernes på Grønland," *Norge tidskrift om vårt land* 3, no. 23 (February 1927).

32. Ad. S. Jensen, "Moskusoksen paa Grønland og dens Fremtid," Report of the 18. Scandinavian Naturalist Congress in Copenhagen, pages 2–3, RA/S-3418/F/Fi/L0001/0002 "Grönland, Fredningsbestemmelser Moskus-okser m.m.," Handelsdepartementet, Industrikontoret 1907–1946, National Archives of Norway.

33. Hoel, "Overføring av moskusokser til Svalbard."

34. Newspaper clipping, "Moskusdyrer på Dovrefjell trives godt," ND [1932], NP Box 116, folder "Moskusokser. Overføring til Norge," National Archives of Norway.

35. Hoel corresponded several times with persons involved with the American Committee for International Wild Life Protection, which had commissioned a major treatise on muskoxen—Elisabeth Hone, *The Present Status of the Muskox in Arctic North America and Greenland* (Cambridge, MA: American Committee for International Wild Life Protection, 1934)—to make this argument. See, for example: Letter from Adolf Hoel to Professor Agersborg, dated April 16, 1934; and letter from Adolf Hoel to Betty Hone, dated April 25, 1934, NP box 116, folder "Moskusokser— Fredning, zoology, avhandlinger, etc.," National Archives of Norway.

36. Adolf Hoel interviewed in "Moskusokser på Svalbard vil bety en verdifull forøkelse av ø-gruppens viltbestand," *Aftenposten*, August 27, 1929, 1.

37. Odd Berset, "Den selsomme moskus og hvordan den kom til Svalbard," *Aftenposten*, June 1, 1957.

38. Spesialtelegram til Aftenposten, "Vårt høifjell blir påny 'befolket' med moskusdyr," *Aftenposten*, October 10, 1932, 1.

39. O. Olstad and P. Tuff, "Innplanting av moskusokser på Dovrefjell," in *Årsmelding om Det Norkse Skogvesen* (Oslo: Skogdirektøren, 1942), 51.

40. List of permits and payments, "Moskusokser" with "Dovre" hand-written at top, NP Box 116, folder "Moskusokser. Overføring til Norge," National Archives of Norway.

41. Spesialtelegram til Aftenposten, "Vårt høifjell blir påny 'befolket' med moskusdyr," *Aftenposten*, October 10, 1932, 1.

42. Vålåsjø station, "Moskusdyrene på Dovrefjell trives udmerket," *Aftenposten*, November 24, 1932, 9.

43. Adolf Hoel, "Moskusokser til Svalbard," *Norsk Jæger- og Fisker Forenings Tidsskrift* 58 (1929): 328.

44. Olstad and Tuff, "Innplanting av moskusokser på Dovrefjell," 52.

45. "Innkjøp av moskusdyr," RA/S-6087/D/Da/Dab/L0090/0004, Norwegian National Archive.

46. Odd Løno, *Transplantation of Muskox in Europe and North-America*, Meddelelser no. 84 (Oslo: Norsk Polarinstitutt, 1960), 9–10.

47. Letter from Anders Orvin and J. Giever to Norsk Telegramsbyrå, dated September 2, 1947, NP Box 244, folder 545, "Om innførsel av moskus til Svalbard og fastlandsnorge," National Archives of Norway. Also publicly published as "Moskuskalver sloppet på Dovre igjen," *Aftenposten*, September 4, 1947, 1.

48. For example, fear was identified as the driving factor for locals who opposed a recent European bison reintroduction project in Germany. See Stephen E. Decker et al., "The Return of the King or Bringing Snails to the Garden? The Human Dimensions of a Proposed Restoration of European Bison (*Bison bonasus*) in Germany," *Restoration Ecology* 18, no. 1 (2010). In "Chemical Unknowns," Egan also discusses how fear of environmental contaminants is integral to modern environmental decision-making processes.

49. Nicolette, "Moskus på Dovre," *Aftenposten*, October 24, 1932, 4.

50. Alfred Hoel to Kongelige Landbruksdepartement, December 10, 1932, NP box 116, folder "Moskusokser. Overføring til Norge," National Archives of Norway.

51. Spesialtelegram til Aftenposten, "Vårt høifjell blir påny 'befolket' med moskusdyr," *Aftenposten*, October 10, 1932, 1.

52. Some examples from *Aftenposten* include "Moskusdyrene på Dovrefjell er meget fredsommelige," August 28, 1942, 1; and "To mosusokser kommet til Grøadalen på Nordmøre," June 18, 1949, 1.

53. "Vanlige stut er farligere," *Aftenposten*, July 20, 1950.

54. Letter from Jac Skylstad to Adolf Hoel, dated September 21, 1937, NP box 116, folder "Moskusokser. Overføring til Norge," National Archives of Norway. In the same archive, the veterinarian Olstad confirmed in a meeting on September 16, 1937, with Hoel and others at the Polar Institute office that there had been three recent attacks.

55. "Møte med moskusokseflokk ur Drivdalen," *Aftenposten*, July 13, 1954, 2.

56. "Ville fotografere moskusokse, ble tatt på hornene," *Aftenposten*, September 2, 1963, 1.

57. "Moskus i harnisk," *Aftenposten*, September 18, 1963, 1.

58. This retelling of the story is based on two articles: "Eldre mann ble drept av en moskusokse i Oppdal," *Aftenposten*, July 23, 1964, morning edition, 1 and 17; and "Vær på vakt mot enslige moskusdyr," *Aftenposten*, July 23, 1964, afternoon edition, 1–2.

59. Letter printed in "122 oppdøler: Moskusen vekk, eller den avlives!" *Aftenposten*, August 11, 1964, morning edition, 1 and 6.

60. "122 oppdøler: Moskusen vekk, eller den avlives!" *Aftenposten*, August 11, 1964, morning edition, 6.

61. "Stemningen i Åmotdalen er fremdeles amper," *Aftenposten*, August 13, 1964, evening edition, 6.

62. "122 oppdøler: Moskusen vekk, eller den avlives!" *Aftenposten*, August 11, 1964, morning edition, 6.

63. "Moskusen beskyttet av loven," *Aftenposten*, August 28, 1964, morning edition, 1 and 31; and "Misnøye i Oppdal med avgjørelsen i moskus-saken," *Aftenposten*, August 28, 1964, evening edition, 1–2.

64. "Moskusbestand på 18–30 dyr," *Aftenposten*, August 11, 1964, afternoon edition, 1 and 7.

65. "Moskusdyr som viser seg i Engan vil bli skutt," *Aftenposten*, September 10, 1964, evening edition, 8.

66. Animals acting outside of human expectations in restoration projects can compromise the expected outcome by pressuring power networks to change, as in the case of otters reintroduced in Missouri, detailed in T. L. Goedeke and S. Rikoon, "Otters as Actors: Scientific Controversy, Dynamism of Networks, and the Implications of Power in Ecological Restoration," *Social Studies of Science* 38, no. 1 (2008).

67. "Vilda myskoxar i Härjedalen. Varning: De kan gå till anfall!" *Östersunds-Posten*, September 1, 1971, 1.

68. Björn Berglung, "Hel hjord på väg in i Sverige," *Dagens Nyheter*, September 5, 1971, 1; and Torolf Byfalt, "Världssensation faktum Myskoxarna i Lofsdalen," *Östersunds-Posten*, October 11, 1971, 1.

69. Björn Berglung, "Hel hjord på väg in i Sverige," *Dagens Nyheter*, September 5, 1971, 1.

70. Krystyna Pieniezny, "Myskoxar på svensk mark," *Svenska Dagbladet*, September 8, 1971, 19.

71. "Myskoxarna vid sportstuga," *Östersunds-Posten*, September 7, 1971, 1; and cartoon in *Östersunds-Posten*, September 3, 1971, 3.

72. "Moskusoksene blir i Sverige," *Aftenposten*, January 14, 1972, 14.

73. Olof Ternström, "Rapport över myskoxarnas revirbeteende under åren 1971–1977," February 10, 1976, folder 270-2665-76, Naturvårdsverket archive.

74. Letter from N. G. Lundh, Fjällnäs Fjällhotell och Turistgård, to Statens naturvårdsverk, May 4, 1976, folder 270-2665-76, Naturvårdsverket archive.

75. Letter from Bengt Andersson on behalf of Tännäs sameby to Länstryelsen i Jämtlands län, August 6, 1976, folder 270-2665-76, Naturvårdsverket archive.

76. This kind of self-interest when dealing with wildlife damage is certainly not unique to this case or to rewilding. The history of the wolf in North America, where wolves have been hunted almost to extinction as well, also shows how conflict over wolf predation of species that humans desire to eat has led to systematic destruction of the animal population. Efforts to reintroduce wolves have been met with significant resistance by local farming and ranching populations, as noted in the conclusion of Jon T. Coleman, *Vicious: Wolves and Men in America* (New Haven, CT: Yale University Press, 2004), 225–235.

77. Typed notes from the meeting were filed with Naturvårdsverket as "Anteckningar från symposium den 2 September 1976," folder 270-2665-76, Naturvårdsverket archive. All descriptions of the discussion at the meeting come from this source.

78. "Deltagarförteckning," folder 270-2665-76, Naturvårdsverket archive.

79. "Promemoria om vård och förvaltning av myskoxarna i Sverige," January 6, 1977, folder 270-2665-76, Naturvårdsverket archive.

80. Letter from Eric Skogland and Pär Hansson, Lantbruksstyrelsen, to Statens Nautvårdsverk, October 24, 1978; letter from Lennart Hjelm and Bo Lidhe, Sveriges Lantburksuniversitet, to Statens naturvårdsverk, October 27, 1978; and letter from Carl Edelstam and Carl-Fredrik Lundevall, Naturhistoriska Riksmuseet, to Statens naturvårdsverk, October 26, 1978, folder 270-2665-76, Naturvårdsverket archive.

81. Letter from Svenska Samernas Riksförbund (SSR) to Statens naturvårdsverk, October 6, 1978, folder 270-2665-76, Naturvårdsverket archive.

82. Letter from Lars Calleberg and Olof Ternström, Jämtlands Länsstyrelsen, to Statens naturvårdsverk, November 3, 1978, folder 270-2665-76, Naturvårdsverket archive.

83. Letter from Iwan Lundberg, Lantbruksnämnden i Jämtlands Län, to Statens naturvårdsverk, September 28, 1978, folder 270-2665-76, Naturvårdsverket archive.

84. Swedish Post Office, Stamps and Philatelic Service, *Fjällvärld* (Mountain World) first-day-issue card, March 27, 1984, author's personal collection.

85. Quote by Nils Lundh, in Charlotte Permell, "Vem dödar myskox-arna?," *Expressen*, April 13, 1994.

86. Sveriges Riksdag, Motion 1990/91:Jo759, "Rädda myskoxstammen."

87. Sveriges Riksdag, Betänkande 1990/91:JoU30, "Miljöpolitiken."

88. Sveriges Riksdag, Motion 1999/2000:MJ763, "Myskoxar i Sverige."

89. Jørgensen, "Muskox in a Box."

90. For example, according to articles in *Aftenposten* in 1973–1974, a muskox was shot in Luster, about 100 km from Dovre (May 3, 1973); an old bull was shot in the valley of Haverdalsssetrene after it had been relocated higher in the mountains three times and each time had returned to that valley (October 25, 1973); and a twelve-year-old from Oppdal almost collided with an animal while riding her bike to the bus stop (June 17, 1974).

91. Per 1980 coverage in *Aftenposten*, for example, a muskox caused a commotion when it showed up in the streets of Åndalsnes and was shot (August 4), and another showed up near the center of Lillehammer 125 km away and was anesthetized and taken to Skåne zoo (October 17 and November 17).

92. Fylkesmannen i Sør-Trøndelag, *Forvaltningsplan for Moskus på Dovre*, report 2/96, May 1996 (Trondheim: Fylkesmannen i Trøndelag, Miljøvernavdelingen), 6.

93. Fylkesmannen i Sør-Trøndelag, *Forvaltningsplan for Moskus på Dovre*, 1996, 5.

94. An overview and detailed map are provided in the 1996 plan. It is obvious that the border on the southeast, which is a perfectly straight line on the map, does not match the topography.

95. Fylkesmannen i Sør-Trøndelag, *Forvaltningsplan for moskus på Dovre*, 1996, 6.

96. Fylkesmannen i Sør-Trøndelag, *Forvaltningsplan for moskus på Dovre*, 1996, 7.

97. Fylkesmannen i Sør-Trøndelag, *Forvaltningsplan for moskus på Dovre*, 1996, 7.

98. Fylkesmannen i Sør-Trøndelag, *Forvaltningsplan for moskusstammen på Dovrefjell*, report 1/2006, February 2006 (Trondheim: Fylkesmannen i Trøndelag, Miljøvernavdelingen).

99. Fylkesmannen i Sør-Trøndelag, *Forvaltningsplan for moskusbestanden på Dovrefjell*, report 4/2017, December 2017 (Trondheim: Fylkesmannen i Trøndelag, Miljøvernavdelingen).

100. Randi Bakke, "Er vi tjent med moskus i Norge?," *Aftenposten*, October 20, 1975, 12.

101. Kretz, "Hope in Environmental Philosophy," 926.

4 Resurrecting

1. Stewart Brand, "'Bringing Back the Passenger Pigeon': Meeting Convened at Harvard Medical School in Boston on Feb. 8, 2012," Long Now Foundation website, http://longnow.org/revive/passenger-pigeon-workshop/.

2. For a discussion of Brand and the whole earth idea, see William Bryant, "Whole System, Whole Earth: The Convergence of Ecology and Technology in Twentieth Century American Culture," PhD diss., University of Iowa, May 2006.

3. For the philosophical position of the Long Now Foundation on long-term thinking, see Stewart Brand, *Clock of the Long Now: Time and Responsibility* (New York: Basic Books, 1999)

4. Long Now Foundation, "Revive & Restore," http://longnow.org/revive/.

5. Brand, "'Bringing Back the Passenger Pigeon.'" We should note that ethics was not a particularly prominent part of this discussion. For an overview of the ethical arguments for and against de-extinction, see Ronald Sandler, "The Ethics of Reviving Long Extinct Species," *Conservation Biology* 28, no. 2 (2013).

6. Greenberg, *A Feathered River across the Sky*.

7. See Heise, *Imagining Extinction*.

8. Elisabeth Kübler-Ross, *On Death and Dying* (New York: Macmillan, 1969).

9. Elisabeth Kübler-Ross and David Kessler, *On Grief and Grieving: Finding the Meaning of Grief through the Five Stages of Loss* (New York: Scribner, 2005).

10. For good overviews of the criticisms of and problems with the five stages in counseling situations, see Ruth David Konigsberg, *The Truth about Grief: The Myth of Its Five Stages and the New Science of Loss* (New York: Simon and Schuster, 2011); and Margaret Stroebe, Henk Schut, and Kathrin Boerner, "Cautioning Health-Care Professionals: Bereaved Persons Are Misguided through the Stages of Grief," *OMEGA—Journal of Death and Dying* 74, no. 4 (2017).

11. See Neil Small, "Theories of Grief: A Critical Review," in Jennifer L. Hockey, Jeanne Katz, and Neil Small, eds., *Grief, Mourning and Death Ritual,* (London: Open University Press, 2001).

12. For the foundational discussion of melancholy, see Sigmund Freud, "Trauer under Melancholie" [Mourning and Melancholia], *Internationale Zeitschrift für Ärztliche Psychoanalyse* 4 (1917).

13. Kathy Charmaz and Melinda J. Milligan, "Grief," in *Handbook of the Sociology of Emotions*, ed. J. E. Stets and J. H. Turner (Boston: Springer, 2006). Thinking through and facing the end of nature in the Anthropocene has been portrayed in literature as a grief-stricken condition, potentially leading even to never-ending melancholia: see Margaret Ronda, "Mourning and Melancholia in the Anthropocene," *Post 45*, October 6, 2013, http://post45.research.yale.edu/2013/06/mourning-and-melancholia-in-the-anthropocene/.

14. Mark Catesby, *The Natural History of Carolina, Florida and the Bahama Islands: Containing the Figures of Birds, Beasts, Fishes, Serpents, Insects, and Plants* [...], vol. 1, (London: Printed by the author, 1731), 23.

15. John James Audubon, *Ornithological Biography, or An Account of the Habits of the Birds of the United States of America; Accompanied by Descriptions of the Objects Represented in the Work Entitled The Birds of America, and Interspersed with Delineations of American Scenery and Manners*, vol. 1 (Philadelphia: Judah Dobson, 1831), 320.

16. Audubon, *Ornithological Biography*, vol. 1, 321.

17. Audubon, *Ornithological Biography*, vol. 1, 323.

18. Reactions to the natural sublime in early North America is linked to European settlers' ideas of wilderness and resulted in tensions between preserving the wild and taming it. See William Cronon's classic discussion of this in "The Trouble with Wilderness, or Getting Back to the Wrong Nature," in *Uncommon Ground: Rethinking the Human Place in Nature*, ed. William Cronon (New York: W. W. Norton & Co., 1995).

19. See Ted Steinberg, *Down to Earth: Nature's Role in American History* (Oxford: Oxford University Press, 2002), 59–61, for a concise overview of these developments.

20. For relations between the railroad expansion in the United States and commodity frontiers, see Mark Fiege, *The Republic of Nature: An Environmental History of the United States* (Seattle: University of Washington Press, 2012), chap. 6; and William Cronon, *Nature's Metropolis: Chicago and the Great West* (New York: W. W. Norton, 1991).

21. See Price, *Flight Maps*, chap. 1, for more on hunting practices that developed around the passenger pigeon.

22. Catesby, *The Natural History of Carolina*, 23.

23. Audubon, *Ornithological Biography*, vol. 1, 321.

24. Cotton Mather, untitled, printed in Arlie W. Schorger, "Unpublished Manuscripts by Cotton Mather on the Passenger Pigeon," *Auk* 55, no. 3 (1938), 473.

25. Letter in "Diaries, notebooks, mss. Warren, Oscar Bird. Passenger Pigeon correspondence," University of Minnesota Library, University Archives, http://umedia.lib.umn.edu/node/781683.

26. James Fenimore Cooper, *The Pioneers, or The Sources of the Susquehanna; A Descriptive Tale*, vol. 1 (New York: Charles Wiley, 1823), chap. 22.

27. Audubon, *Ornithological Biography*, vol. 1, 323.

28. "Winter Sports in Northern Louisiana: Shooting Wild Pigeons," *Frank Leslie's Illustrated Newspaper*, February 20, 1875, John B. Cade Library digital collection, http://louisianadigitallibrary.org/islandora/object/subr-hwj%3A376.

29. Audubon, *Ornithological Biography*, vol. 1, 323.

30. Audubon, *Ornithological Biography*, vol. 1, 326. Live passenger pigeons would have been available to purchase because live trapping was done to supply birds for professional shooting contests in the northeastern United States; see Julian P. Hume, "Large-Scale Live Capture of Passenger Pigeons *Ectopistes migratorius* for Sporting Purposes: Overlooked Illustrated Documentation," *Bulletin of the British Ornithologists' Club* 135 (2015).

31. The practice of intentionally releasing species to allow them to establish in new territory was known in the nineteenth and early twentieth centuries as *acclimatization*. European colonizers used acclimatization as a way of remaking newly encountered environments fit their own conceptions of "correct" nature. See Thomas R. Dunlap, "Remaking the Land: The Acclimatization Movement and Anglo Ideas of Nature," *Journal of World History* 8, no. 2 (1997); and Harriet Ritvo, "Going Forth and Multiplying: Animal Acclimatization and Invasion," *Environmental History* 17, no. 2 (2012). This system was practiced within the framework of colonial science and production; see Warwick Anderson, "Climates of Opinion: Acclimatization in Nineteenth-Century France and England," *Victorian Studies* 35, no. 2 (1992); and Michael A. Osborne, "Acclimatizing the World: A History of the Paradigmatic Colonial Science," *Osiris*, 2nd series, 15 (2001).

32. Gregory Smart, *Birds on the British List: Their Title to Enrollment Considered* (London: R. H. Porter, 1886).

33. The correspondence is located at the University of Minnesota Libraries, University Archives, under the local identifier uarc00876-box67-fdr624. The correspondence has been digitized and is available at https://umedia.lib.umn.edu/node/781683.

34. Quoted in W. B. Mershon, *The Passenger Pigeon* (New York: Outing Publishing, 1907), 121.

35. Richard Lydekker, *The Royal Natural History*, vol. 4 (London: Frederick Warne & Co., 1895), 374.

36. Cited in Mershon, *The Passenger Pigeon*, 141–143.

37. "Wild Pigeon Found," *Chicago Tribune*, August 21, 1898, 28.

38. John Burroughs, cited in Mershon, *The Passenger Pigeon*, 180.

39. Mershon, *The Passenger Pigeon*, ix.

40. Mershon, *The Passenger Pigeon*, xii.

41. Mershon, *The Passenger Pigeon*, xii.

42. For example, the letter by C. H. Ames in Mershon, *The Passenger Pigeon*, 173–176.

43. Mershon, *The Passenger Pigeon*, xii.

44. Reprinted in Mershon, *The Passenger Pigeon*, 55.

45. Reprinted in Mershon, *The Passenger Pigeon*, 186.

46. Sullivan Cook, in *Forest and Stream*, March 14, 1903; quoted in Mershon, *The Passenger Pigeon*, 171–172.

47. "Wild Pigeon Found," *Chicago Tribune*, August 21, 1898, 28.

48. For the most up-to-date biography of Hornaday, see Gregory J. Dehler, *The Most Defiant Devil: William Temple Hornaday and His Controversial Crusade to Save American Wildlife* (Charlottesville: University of Virginia Press, 2013). For Hornaday's involvement in the bison conservation program, see also Andrew C. Isenberg, "The Returns of the

Bison: Nostalgia, Profit, and Preservation," *Environmental History* 2, no. 2 (1997). For a critical view of how Hornaday's racism effected his wild-life conservation ideas, see Powell, *Vanishing America*, 63–67.

49. See, for example, Mershon, *The Passenger Pigeon*, xii. Mark V. Barrow Jr. compares the two explicitly in "Teetering on the Brink of Extinction: The Passenger Pigeon, the Bison, and American Zoo Culture in the Late Nineteenth and Early Twentieth Centuries," in *The Ark and Beyond: The Evolution of Zoo and Aquarium Conservation*, ed. Ben A. Minteer, Jane Maienschein, and James P. Collins (Chicago: University of Chicago Press, 2018). Barrow concludes that zoos failed to breed the pigeon the way they did the bison and thus did not contribute to their conservation. The pigeon/buffalo link moves into popular culture as well. For example, I have discussed the way in which the two are paired in the first aired episode of the TV series *Star Trek*: Jørgensen, "Who's the Devil? Species Extinction and Environmentalist Thought in Star Trek," in *Star Trek and History*, ed. Nancy Reagin (New York: Wiley & Sons, 2013).

50. William T. Hornaday, *Our Vanishing Wild Life: Its Extermination and Preservation* (New York: Charles Scribner's Sons, 1913), vii.

51. Hornaday, *Our Vanishing Wild Life*, ix; italics in original.

52. Hornaday, *Our Vanishing Wild Life*, ii

53. Hornaday, *Our Vanishing Wild Life*, 11 and 13; italics in original.

54. Joe Kosack, "Passenger Pigeon: *Ectopistes migratorius*," brochure, Pennsylvania Game Commission, March 2010, http://www.pgc.pa.gov/Wildlife/EndangeredandThreatened/Pages/PassengerPigeon.aspx.

55. See William T. Hornaday, *The Extermination of the American Bison* (Washington, DC: US Government Printing Office, 1889), for a discussion of the various herds, including those in private hands and those in Canada at the time of publication.

56. The National Bison Range of the US Fish and Wildlife Service has created a table showing in detail the major events, including reintroduction measures, that affected the population of the American bison until 1935: https://www.fws.gov/bisonrange/timeline.htm.

57. Zdzisław Pucek et al., *European bison (Bison bonasus): Current State of the Species and Strategy for Its Conservation* (Strasbourg: Council of Europe Publishing, 2004).

58. Ruthven Deane, "Some Notes on the Passenger Pigeon (*Ectopistes Migratorius*) in Confinement," *Auk* 13, no. 13 (1896).

59. Deane, "Some Notes on the Passenger Pigeon," 237.

60. Ruthven Deane, "The Passenger Pigeon (*Ectopistes Migratorius*) in Confinement," *Auk* 25, no. 2 (1908). Of the seven birds given back to Whittaker, four males were still alive in 1907.

61. General information on Whitman is available via the online exhibit "Charles Otis Whitman: His Science, His Special Birds and the Marine Biological Laboratory," University of Chicago Library, https://www .lib.uchicago.edu/collex/exhibits/charles-otis-whitman-his-science-his -special-birds-and-marine-biological-laboratory/. A portrait article on the professor, titled "Lives among Five Hundred Pigeons," ran in the *Chicago Tribune* on October 7, 1900. Many of Charles Darwin's observations about evolution can be linked to his interest in pigeon breeds and pigeon fancying, so using pigeons to investigate inheritance would be a natural choice for later scientists interested in related topics.

62. Whitman provided a memorandum detailing his flock to Ruthven Deane, who published it in "The Passenger Pigeon (*Ectopistes Migratorius*) in Confinement."

63. Special correspondence of the *Transcript*, "The Passenger Pigeon," *Boston Evening Transcript*, May 11, 1901, 23.

64. James H. Fleming, "The Disappearance of the Passenger Pigeon," *Ottawa Naturalist* 20, no. 12 (March 1907): 237.

65. Ella Gilbert Ives, "The Passenger Pigeon: A Plan to Recover a Vanishing Species," *Boston Evening Transcript*, March 16, 1910, 18. Hodge was known for his work with the educational uses of studying nature, as exemplified in his book *Nature Study and Life* (Boston: Ginn & Co., 1902).

66. F. J. Wenninger, "The Passenger Pigeon," *American Midland Naturalist* 1, no. 8 (1910). Additional details are provided in Ives, "The Passenger Pigeon: A Plan to Recover a Vanishing Species." Kuser had withdrawn a previous offer of one hundred dollars for a freshly killed passenger pigeon and replaced it with a $300 reward for information of undisturbed nesting birds: see "Notes and News," *Auk* 27, no. 1 (1910).

67. Ives, "The Passenger Pigeon," 18.

68. C. F. Hodge, "The Passenger Pigeon Investigation," *Auk* 28, no. 1 (1911). The paper had been read at the AOU meeting in November 1910 and was later published in *School Science and Mathematics* 11, no. 4 (April 1911): 356–361.

69. Hodge, "The Passenger Pigeon Investigation," 51. For a discussion of the dilemmas tied to absence of confirmed sightings of a species being equated to extinction, see Dolly Jørgensen, "Presence of Absence, Absence of Presence and Extinction Narratives," in *Nature, Temporality and Environmental Management*, ed. Lesley Head et al. (Abingdon: Routledge, 2017).

70. Hodge, "The Passenger Pigeon Investigation," 51.

71. Hodge, "The Passenger Pigeon Investigation," 52.

72. C. F. Hodge, "A Last Word on the Passenger Pigeon," *Auk* 29, no. 2 (1912). This doesn't mean that there were not more sightings. For example, on April 16, 1911, the *Milwaukee Sentinel* published an article about a passenger pigeon seen in Cottage Grove, Wisconsin, and observed through a small telescope and opera glasses: "Watch Passenger Pigeon," *Milwaukee Sentinel*, April 16, 1911, 6.

73. Hodge, "A Last Word," 174.

74. Hodge, "A Last Word," 170. This conclusion was particularly meaningful for Hodge because of his research interest in nature pedagogy.

75. This is the date given in almost all accounts. A text published on September 13 gave the time of death as "2 p.m., August 29, 1914," but this is an error. See John C. French, *The Passenger Pigeon in Pennsylvania* (Altoona, PA: Altoona Tribune Co., 1919), 187.

76. There is some question as to whether Martha had been bred by Whitman or captured as a live bird. French's text says she was one of seventeen birds captured in 1876, but that would have made her far older (thirty-eight years) than any other known wild pigeon, which had life expectancies in the twenties. An alternate story given by William C. Herman is that Martha was actually hatched at the Cincinnati zoo in 1885 by caretaker Sol A. Stephan, who purportedly was able to hatch fourteen passenger pigeons in captivity at the zoo. See William C. Herman, "The Last Passenger Pigeon," *Auk* 65, no. 1 (1948).

77. An alternate story holds that Martha was named in memory of the wife of a friend of Mr. Stephan; see Herman, "The Last Passenger Pigeon."

78. Reprinted from *Washington Times*, "Last Passenger Pigeon Dead in Cincinnati," *Milwaukee Journal*, August 30, 1914, 6.

79. For an analysis of the contemporary comparison of the last of a wildlife species to the last of the Mohicans and the trope of the extinction of the "Indian," see Powell, *Vanishing America*, 119–133 and 138–140.

80. French, *The Passenger Pigeon in Pennsylvania*, 187.

81. R. W. Shufeldt, "Anatomical and Other Notes on the Passenger Pigeon (Ectopistes migratorius) Lately Living in the Cincinnati Zoölogical Gardens," *Auk* 32, no. 1 (1915).

82. Shufeldt, "Anatomical and Other Notes," 38. I have not been able to discover where Martha's heart ended up.

83. Herman, "The Last Passenger Pigeon," 80.

84. "The Passenger Pigeon Extinct," *Schenectady Gazette*, September 16, 1932, 27.

85. "Passenger Pigeon Memorial Site," *Gettysburg Times*, October 8, 1947, 1.

86. I have seen photographs of the current monument and have read the text on plaque from those photographs: http://allenbrowne.blogspot .no/2012/02/pigeon-monument.html.

87. The new brick monument is in Codorus State Park: http://allen browne.blogspot.se/2012/02/pigeon-monument.html.

88. Aldo Leopold, "On a Monument to the Pigeon," in *Silent Wings: A Memorial to the Passenger Pigeon*, ed. Walter E. Scott (Madison: Wisconsin Society for Ornithology, 1947); reprinted with slight modifications in Aldo Leopold, *A Sand County Almanac: And Sketches Here and There* (Oxford: Oxford University Press, 1968).

89. Aldo Leopold, "On a Monument to the Pigeon," 3.

90. Walter E. Scott, "Dedication," in *Silent Wings: A Memorial to the Passenger Pigeon*, ed. Walter E. Scott (Madison: Wisconsin Society for Ornithology, 1947), 1.

91. H. T. Jackson, "Attitude in Conservation," in *Silent Wings: A Memorial to the Passenger Pigeon*, ed. Walter E. Scott (Madison: Wisconsin Society for Ornithology, 1947), 22 and 24.

92. Office of Communications, Department of the Interior, "A Passing in Cincinnati, September 1, 1914" (Washington, DC: US Government Printing Office, 1976).

93. When Martha died, her body was shipped by rail rather than air.

94. Richard Kurin, *The Smithsonian's History of America in 101 Objects* (New York: Penguin Press, 2013), chap. 28.

95. The event videos and report are available at http://longnow.org/revive/events/tedxdeextinction.

96. Archer led a project at the Australian Museum that attempted to reconstruct the genome of the thylacine, also known as the Tasmanian tiger, from 1999 to 2005. This carnivorous marsupial had been hunted heavily by settlers in Tasmania, who perceived it as a threat to livestock. The last known thylacine died in a zoo in Hobart in 1936. The project team successfully recovered and replicated some DNA segments from a thylacine pouch pup specimen. The project was officially discontinued in 2005, primarily because of the lack of quality specimens from which to recover DNA. The intended outcome of the project had been to clone an embryo that then could be implanted into a Tasmanian

devil, the closest living relative. The project produced a glossy brochure called "Thylacine Project," a copy of which I have seen in the National Archives of Australia. The project is discussed in Amy L. Fletcher, "Bring 'em Back Alive: Taming the Tasmanian Tiger Cloning Project," *Technology in Society* 30 (2008); and Turner, "Open-Ended Stories."

97. Examples include Hannah Waters, "The Narcissism of De-extinction," *Scientific American* blog, March 15, 2013, http://blogs.scientificamerican.com/culturing-science/2013/03/15/deextinction; Virginia Gewin, "Laws Lag Behind Science in De-extinction Debate," *Discover Magazine*, June 5, 2013, blogs.discovermagazine.com/crux/2013/06/05/laws-lag-behind-science-in-de-extinction-debate; and David Biello, "Efforts to Resuscitate Extinct Species May Spawn a New Era of the Hybrid," *Scientific American*, March 26, 2013, https://www.scientificamerican.com/article/lost-species-revived-from-dna-and-restored-to-nature/.

98. See Helena Siipi and Leonard Finkelman, "The Extinction and De-extinction of Species," *Philosophy and Technology* 30, no. 4 (2017), for an overview of ways to understand the outcomes of resurrection biology from a philosophical standpoint.

99. Ben Novak, "Reflections on Martha's Centennial," Long Now Foundation website, September 1, 2014, http://longnow.org/revive/of-martha-the-last-passenger-pigeon/.

100. Novak, "Reflections on Martha's Centennial."

101. Stewart Brand, "Why Revive Extinct Species," http://longnow.org/revive/what-we-do/why-revive-extinct-species/.

102. As of October 2018, Novak has created several batches of genetically modified (chimera) rock pigeons in the first stages of making pigeons with "correct" passenger pigeon genetic traits. See Amy Dockser Marcus, "Meet the Scientists Bringing Extinct Species Back from the Dead," *Wall Street Journal*, October 11, 2018, https://www.wsj.com/articles/meet-the-scientists-bringing-extinct-species-back-from-the-dead-1539093600.

103. Marjorie Adams, "The Passenger Pigeon," *Victoria Advocate*, September 11, 1970, 44.

104. Amy L. Fletcher, "Genuine Fakes: Cloning Extinct Species as Science and Spectacle," *Politics and the Life Science* 20, no. 1 (2010).

105. Ursula K. Heise, "Lost Dogs, Last Birds, and Listed Species: Cultures of Extinction," *Configurations* 18 (2010).

5 Remembering

1. Information in this paragraph is taken from the JJ1/Bruno exhibit at the Museum Mensch und Natur, which I visited. I have elected to call the bear Bruno because that is the name he is given in the international media coverage. In some news coverage in Austria, he was also referred to as Beppo.

2. Maurice Halbwachs, *On Collective Memory* (Chicago: University of Chicago Press, 1992).

3. Jan Assmann, "Communicative and Cultural Memory," in *Cultural Memory Studies: An International and Interdisciplinary Handbook*, ed. A. Eril and A. Nünning (Berlin: Walter de Gruyter, 2008).

4. Harald Welzer, "Communicative Memory," in *Cultural Memory Studies: An International and Interdisciplinary Handbook*, ed. A. Eril and A. Nünning (Berlin: Walter de Gruyter, 2008), 285.

5. Welzer, "Communicative Memory," 295.

6. As quoted in Marek Tamm, "Beyond History and Memory: New Perspectives in Memory Studies," *History Compass* 11, no. 6 (2013): 464.

7. See Simon Schama, *Landscape and Memory* (Toronto: Random House, 1995), for a sweeping examination of the cultural mental construction of environments.

8. Daniel Pauly, "Anecdotes and the Shifting Baseline Syndrome of Fisheries," *Trends in Ecology & Evolution* 10, no. 10 (1995).

9. S. K. Papworth et al., "Evidence for Shifting Baseline Syndrome in Conservation," *Conservation Letters* 2 (2009); Frans Vera, "The Shifting Baseline Syndrome in Restoration Ecology," in *Restoration and History: The Search for a Usable Environmental Past*, ed. Marcus Hall (New York:

Routledge, 2009); and Miguel Clavero, "Shifting Baselines and the Conservation of Non-native Species," *Conservation Biology* 28, no. 5 (2014).

10. For a foundational examination of the past in the present, see David Lowenthal, *The Past Is a Foreign Country* (Cambridge: Cambridge University Press, 1985), and an insightful extended review essay of the same in C. Vann Woodward, "Review of *The Past Is a Foreign Country*," *History and Theory* 26, no. 3 (1987).

11. Tom Bristow, "Memory," *Environmental Humanities* 5 (2014).

12. William Hornaday, who I discussed in chapter 4, argued that taxidermy displays should be designed to show moral values: see Hanna Rose Shell, "Skin Deep: Taxidermy, Embodiment and Extinction in W. T. Hornaday's Buffalo Group," *Proceedings of the California Academy of Sciences* 55, S1, no. 5: 100–102.

13. Jeffrey K. Stine had some good thoughts about the potential integration of environmental history in a wide variety of museum settings in "Placing Environmental History on Display," *Environmental History* 7, no. 4 (2002). It is unfortunate that his proposals for exhibition reviews in the journal, a website dedicated to listing environmental-history-related exhibitions, and an exhibition prize have never been adopted.

14. Information on the brown bear's near extinction and then early reintroduction efforts is taken from Wilko Graf von Hardenberg, "Another Way to Preserve: Hunting Bans, Biosecurity and the Brown Bear in Italy, 1930–60," in *The Nature State: Rethinking the History of Conservation*, ed. Wilko Graf von Hardenberg et al. (Abingdon: Routledge, 2017).

15. Information about the LIFE Ursus project is available at http://www.pnab.it/natura-e-territorio/orso/life-ursus.html. The follow-up EU LIFE project Arctos ran from 2010 to 2014 to continue work on brown bear conservation in the Alps and Apennines (http://www.life-arctos.it/).

16. Neither Germany as a nation or the state of Bavaria had a brown bear management plan at that time because they had no intention of having brown bears until then. "Managementplan Braunbären in Bayern" was adopted in 2007 as a response to the incident.

17. A good summary of Bruno's relationship to the 2006 World Cup is found in Brian Blickenstaff, "Remembering Bruno, the Problem Bear that Overshadowed the 2006 World Cup," *Vice Sports*, July 13, 2016, https://sports.vice.com/en_us/article/aebez5/remembering-bruno-the -problem-bear-that-overshadowed-the-2006-world-cup.

18. "Nach 170 Jahren wider ein Bär in Bayerns Wäldern," *Bild*, May 22, 2006. On display in the Museum Mensch und Natur exhibit.

19. Information on Rewilding Europe and its programs can be found at https://www.rewildingeurope.com.

20. Natasha Seegert, "Queer Beasts: Ursine Punctures in Domesticity," *Environmental Communication: A Journal of Nature and Culture* 8, no. 1 (2014), https://doi.org/10.1080/17524032.2013.798345. From a scientific wildlife management standpoint, the Bruno incident has been used to advocate more multinational cooperation on reintroduction and classification of nuisance animals; see Tatjana Rosen and Alistair Bath, "Transboundary Management of Large Carnivores in Europe: From Incident to Opportunity," *Conservation Letters* 2 (2009). A short editorial in the *Indoor Built Environment* journal published immediately after the incident drew the conclusion that "maybe the story of Bruno is not just about a lonely bear. Conservation is not just to prevent the extinction of a species whose presence has become inconvenient. It is part of a larger struggle for the ultimate preservation of ourselves." See John H. Lange, "Bruno the Bear," *Indoor Built Environment* 15, no. 5 (2006): 391–392.

21. Shell, "Skin Deep," 104.

22. Rachel Poliquin, *The Breathless Zoo: Taxidermy and the Cultures of Longing* (University Park: Pennsylvania State University Press, 2012), 7.

23. Petra Fohrmann, *Bruno alias JJ1. Reisetagebuch eines Bären* (Swisttal: Fohrmann, 2006), 20–21, 36–37.

24. Fohrmann, *Bruno alias JJ1*, 74–75.

25. Fohrmann, *Bruno alias JJ1*, 98–99.

26. Heinz Vogel, *Das abenteuerliche Leben des JJ1 alias Bruno alias Beppo. Ein Bilderbuch für Erwachsene und Kinder* (Hohenems: Bucher, 2013).

27. The grave marker is even entered into the Find a Grave database (www.findagrave.com). Photographs of the two monuments are shown in Brigitte Warenski, "Bär Bruno wurde total vermenschlicht," *Tiroler Tageszeitung*, June 26, 2016.

28. Sabine Dobel, "Zehn Jahre nach 'Bruno': Kanne in neuer Bär kommen?" *Augsburger Allgemeine*, May 19, 2016; and Brigitte Warenski, "Bär Bruno wurde total vermenschlicht," *Tiroler Tageszeitung*, June 26, 2016. Other examples include "Geliebtm gefürchtet—ausgestopft," *Bayerische Staatszeitung*, May 19, 2016; and "Zehn Jahre nach seinem Tod," *BR24*, June 26, 2016.

29. "Memorials," Todd McGrain Sculpture, http://www.toddmcgrain .com/memorials.

30. For an exploration of the function of monuments to extinction, see Dolly Jørgensen, "After None: Memorialising Animal Species Extinction through Monuments," in *Animals Count: How Population Size Matters in Animal-Human Relations*, ed. Nancy Cushing and Jodi Frawley (Abingdon: Routledge, 2018).

31. The artistic designer of *Martha, the Last Passenger Pigeon* was John A. Ruthven. The mural itself was painted with the help of youth apprentices.

32. Greenberg, *A Feathered River across the Sky*, xii–xiii.

33. Fuller, *Passenger Pigeon*, 10.

34. Avery, *A Message from Martha*, preface.

35. The paradox of wildlife tourism that includes eating the meat of the animals targeted by the tourist experience is explored in Muchazondida Mkono, "'Eating the Animals You Come to See': Tourists' Meat-Eating Discourses in Online Communicative Texts," in *Animals and Tourism: Understanding Diverse Relationships*, ed. Kevin Markwell (Bristol: Channel View Publications, 2015).

36. It is interesting to note that a fountain with a muskox statue had been proposed by the Oppdal Tourist and Outdoor Life Board for Oppdal in 1957, but it was never installed. The proposal with pictures was discussed in "I Dagens Löp," *Aftenposten*, February 5, 1957.

37. This is part of the more extended history of moving muskoxen to Norway. For discussion of the context of the Svalbard introduction (I call it an *introduction* because muskoxen are not believed to have lived in Svalbard in the past), see Roberts and Jørgensen, "Animals as Instruments of Norwegian Imperial Authority in the Interwar Arctic"; and Jørgensen, "Muskox in a Box."

38. I visited the Biological Museum in 2012. In 2017, it was closed indefinitely.

39. I've explored this notion of declaring belonging through placement in the case of animals depicted on Nordic premodern maps: Dolly Jørgensen, "Beastly Belonging in the Premodern North," in *Visions of North in Premodern Europe*, ed. Dolly Jørgensen and Virginia Langum (Turnhout: Brepols, 2018).

40. Direktoratet for Naturforvalting, "Stor dødelighet på moskus," August 28, 2012, http://sdc.sft.no/no/Nyheter/Nyheter/Nyhetsarkiv/2012/8/Stor-dodelighet-pa-moskus/; and Vidar Heitkøtter, "73 moskuser døde," *Gudbransdølen Dagningen*, https://www.gd.no/nyheter/73-moskuser-dode/s/1-934610-6417050. The disease has been identified as a zoonotic bacteria that moved from domesticated sheep into the muskox population: see Kjell Handeland et al., "*Mycoplasma ovipneumoniae*: A Primary Cause of Severe Pneumonia Epizootics in the Norwegian Muskox (*Ovibos moschatus*) Population," *PLoS One* 9, no. 9 (2014): e106116.

41. I visited the exhibit in June 2013.

42. J. W. H. Conroy and A. C. Kitchener, *The Eurasian Beaver (Castor fiber) in Scotland: A Review of the Literature and Historical Evidence*, Scottish Natural Heritage Commission Report No. 49 (1996); and David W. Macdonald et al., "Reintroducing the Beaver (*Castor fiber*) to Scotland: A Protocol for Identifying and Assessing Suitable Release Sites," *Animal Conservation* 3 (2000).

43. J. Gurnell et al., *The Feasibility and Acceptability of Reintroducing the European Beaver to England*, Natural England Commissioned Report NECR002 (Peterborough: Natural England and the People's Trust for Endangered Species, 2009).

44. Scottish National Heritage, *Re-introduction of the European beaver to Scotland: A Public Consultation* (Perth: Scottish Natural Heritage, 1998), 9.

45. R. Edwards, "Beavers Finally Allowed to Return ... after 400 Years," *Herald Scotland*, May 24, 2008, http://www.heraldscotland.com/beavers-finally-allowed-to-return-after-400-years-1.828770.

6 Reconnecting

1. This statement appears only in the version published as Aldo Leopold, *A Sand County Almanac: And Sketches Here and There* (Oxford: Oxford University Press, 1968), 112; italics in original. In the *Silent Wings* version (as well as the *Sand County Almanac*), there is a statement earlier in the text that reads, "For one species to mourn the death of another is a new thing under the sun." The additional statement about love at the very end of the piece plays off that earlier one.

2. Leopold, "On a Monument to the Pigeon," 3–4.

3. Philip Cafaro and Richard Primack, "Species Extinction Is a Great Moral Wrong," *Biological Conservation* 170 (2014): 2.

4. "Bäverkärlek," *Dagens Nyheter*, October 24, 1935, newspaper clipping, folder E.5, "Natur- och Djurskydd 1935–37," Jamtli archive.

5. Telegram from Bäverjensen to Festin, October 13, 1935, folder C 32.2, "Bäverinplanteringen, 1935," Jamtli archive. The telegram was also reported in the newspaper: "En bäverhälsning," unlabeled newspaper clipping in Jamtli archive, folder G1:1.

6. E. O. Wilson, *Biophilia* (Cambridge, MA: Harvard University Press, 1984).

7. Leopold, "On a Monument to the Pigeon," 3.

8. Barrow, *Nature's Ghosts*; and Jones, *Empire of Extinction*.

9. For the bison, see Andrew C. Isenberg, *The Destruction of the Bison: An Environmental History, 1750–1920* (Cambridge: Cambridge University Press, 2000); for the thylacine (a.k.a. Tasmanian tiger), see Robert Paddle, *The Last Tasmanian Tiger: The History and Extinction of the Thylacine* (Cambridge: Cambridge University Press, 2000).

References

Archival Sources Consulted

Aftenposten

Digital archive.

Jamtli archives, Östersund, Sweden

Folder C32.2 Bäverinplanteringen, 1920–1943.

Folder E.1 Natur- och Djurskydd 1927–1929.

Folder E.2 Natur- och Djurskydd 1920–1926.

Folder E.4 Natur- och Djurskydd 1934–1935.

Folder E.5 Natur- och Djurskydd 1935–1937.

Folder F8FB:1 Kultureminnesvård, Naturvård, Bäverinplantering.

Folder G1:1 Foton, Bäverinplantering Jämtlands län Bildarkiv.

National Archives of Norway (Oslo)

RA/S-1245/D/Dc/L0003 Viltstyrets dok. Fra okkupajonstiden.

RA/PA-1637/D/L0002/0003 Moskusokser, Grønland.

RA/S-1260/D/Dx/L0005/0008 Adolf Hoel, anvisninger.

RA/S-2263/F/Fa/Fac/L0050/0001 Moskusundersøkelser Dovre.

RA/S-3418/F/Fi/L0001/0002 Grönland, Fredningsbestemmelser Moskus-okser m.m., Handelsdepartementet, Industrikontoret 1907–1946.

RA/S-4247/D/ L0095 Moskus §34, Landbruksdepartementet, Kontorer for viltstell, jakt og fangst.

RA/S-6087/D/Da/Dab/L0089 Bever, Direktoratet for vilt og ferkvannsfisk.

RA/S-6087/D/Da/Dab/L0090 Moskusdyr, Direktoratet for vilt og friskvannsfisk.

RA/S-6087/D/Da/Dab/L0090/0004 Innkjøp av moskusdyr.

National Archives of Norway (Tromsø)

Norsk Polarinstitutt box 116.

Norsk Polarinstitutt box 139.

Norsk Polarinstitutt box 244.

Naturvårdsverket Archive, Stockholm, Sweden

Folders 270-2665-76, 273-4425-76, 270-3289-76, 270-1976-77, 260-2135-77.

Nordiska Museet, Stockholm, Sweden

Skansen Zoologiska trädgården innkommende korrespondens.

Norwegian Polar Institute Archive, Tromsø, Norway

Småskrift section.

Sveriges Riksdag

Digital portal.

University of Minnesota Library, University Archives

Passenger pigeon correspondence (http://umedia.lib.umn.edu/node/78 1683).

Printed Primary Sources

Anderson, Axel. "Då bävern återbördades till Västerbotten." *Västerbotten: Västerbottens läns hembygdsförenings årsbok* 5 (1924–1925): 282–286.

Anonymous. "Notes and News." *Auk* 27, no. 1 (1910): 112.

Arbman, Sven. "När bäfvern återinfördes i Bjurälfen." *Svenska Jägareförbundets Tidskrift* 60 (1922): 274–280.

Audubon, John James. *The Birds of America from Original Drawings*, vol. 1. London: Published by the author, 1827–1830.

Audubon, John James. *Ornithological Biography, or An Account of the Habits of the Birds of the United States of America; Accompanied by Descriptions of the Objects Represented in the Work Entitled* The Birds of America, *and Interspersed with Delineations of American Scenery and Manners*, vol. 1. Philadelphia: Judah Dobson, 1831.

Avery, Mark. *A Message from Martha: The Extinction of the Passenger Pigeon and Its Relevance Today*. New York: Bloomsbury, 2014.

Behm, Alarik. "Några ord om naturskydd." *Jämten* 14 (1920): 40–51.

Behm, Alarik. "Naturskydd, särskilt i Jämtland." *Sveriges Natur* (1921): 121–130.

Behm, Alarik. *Nordiska Däggdjur: 177 bilder från Skansen*. Uppsala: J. A. Lindblads Bokförlags Aktiebolag, 1922.

Brand, Stewart. *Clock of the Long Now: Time and Responsibility*. New York: Basic Books, 1999.

Catesby, Mark. *The Natural History of Carolina, Florida and the Bahama Islands: Containing the Figures of Birds, Beasts, Fishes, Serpents, Insects, and Plants: Particularly, the Forest-Trees, Shrubs, and Other Plants, not Hitherto Described, or Very Incorrectly Figured by Authors: Together with Their Descriptions in English and French: To Which, Are Added Observations on the Air, Soil, and Waters: With Remarks upon Agriculture, Grain, Pulse, Roots, &c.: To the Whole, Is Prefixed a New and Correct Map of the Countries Treated Of*, vol. 1. London: Printed by the author, 1731.

Collett, Robert. "Bæveren i Norge, dens Udbredelse og Levemaade (1896)." *Bergens museums aarbog* (1897): article 1, 1–127.

Collett, Robert. "Meddelelser om Norges Pattedyr i Aarene 1876–1881." *Nyt Magazin for Naturvidenskaberne* 27 (1882): 226–230.

Collett, Robert. "Om Bæveren (*Castor fiber*), og dens Udbredelse i Norge fordum og nu." *Nyt Magazin for Naturvidenskaberne* 28 (1883): 11–45.

Conroy, J. W. H., and A. C. Kitchener. *The Eurasian Beaver (Castor fiber) in Scotland: A Review of the Literature and Historical Evidence.* Scottish Natural Heritage Commission Report No. 49 (1996).

Cooper, James Fenimore. *The Pioneers, or The Sources of the Susquehanna; A Descriptive Tale*, vol. 1. New York: Charles Wiley, 1823.

de Blainville, M. H. "Sur plusieurs espéces d'animaux mammiferes, de l'ordre des ruminants." *Bulletin des Sciences, par La Société Philomatique de Paris* (1816): 73–82.

de Charlevoix, Pierre François Xavier. *Journal d'un voyage fait par ordre du roi dans l'Amerique septentrionnale*, vol. 3. Paris: Chez Nyon, 1744.

Deane, Ruthven. "The Passenger Pigeon (*Ectopistes Migratorius*) in Confinement." *Auk* 25, no. 2 (1908): 181–183.

Deane, Ruthven. "Some Notes on the Passenger Pigeon (*Ectopistes Migratorius*) in Confinement." *Auk* 13, no. 13 (1896): 234–237.

Festin, Eric. "Bäverns återinplantering." *Jämten* 16 (1922): 84–91.

Festin, Eric. "Bäverns återinplantering i Jämtland." *Sveriges Natur* 12 (1921): 148.

Festin, Eric. "Fridlysning av Bjurälvdalens karstlandskap och återinplantering av bävern: En samtidig lösning av två viktiga naturskyddsfrågor." *Sveriges Natur* 13 (1922): 32–62.

Festin, Eric. "Jubileumsutställningen och Kulturmässan i Östersund 1920." *Jämten* 14 (1920): 26–31.

Festin, Eric. "Sveriges nya bäverstam." *Sveriges Natur* 19 (1928): 155–156.

Festin, Eric. "Sveriges nya bäverstam: Det första återinplanteringsini-tiativet och dess efterföljare." *Jämtlands läns jaktvårdsförening årsbok* (1928): 17–24.

Fleming, James H. "The Disappearance of the Passenger Pigeon." *Ottawa Naturalist* 20, no. 12 (March 1907): 236–237.

Fohrmann, Petra. *Bruno alias JJ1. Reisetagebuch eines Bären.* Swisttal: Fohrmann, 2006.

Forsslund, Karl-Erik. *Hembygdsvård.* 2 vols. Stockholm: Wahlström & Widstrand, 1914.

French, John C. *The Passenger Pigeon in Pennsylvania.* Altoona, PA: Altoona Tribune Co., 1919.

Fries, Carl. *Bäverland: En bok om bävern och hans verk.* Stockholm: Nordisk Rotogravyr, 1940.

Fuller, Errol. *The Passenger Pigeon.* Princeton, NJ: Princeton University Press, 2015.

Fylkesmannen i Sør-Trøndelag. *Forvaltningsplan for moskusbestanden på Dovrefjell,* report 4/2017, December 2017. Trondheim: Fylkesmannen i Sør-Trøndelag, Mijøvernavdelingen.

Fylkesmannen i Sør-Trøndelag. *Forvaltningsplan for moskus på Dovre,* report 2/96, May 1996. Trondheim: Fylkesmannen i Sør-Trøndelag, Mijøvernavdelingen.

Fylkesmannen i Sør-Trøndelag. *Forvaltningsplan for moskusstammen på Dovrefjell,* report 1/2006, February 2006. Trondheim: Fylkesmannen i Sør-Trøndelag, Mijøvernavdelingen.

Gerald of Wales. *The Itinerary through Wales and the Description of Wales,* edited by W. Llewelyn Williams. London: J. M. Dent, 1912.

Geete, Erik. "Bävern i Sverige och Norge." *Skogen* 16, no. 10 (1929): 298–301; 16, no. 11 (1929): 328–331; 16, no. 12 (1929): 355–358; 16, no. 15 (1929): 417–418; 16, no. 16 (1929): 441–444 (serialized over 5 issues).

Gisler, Nils. "Rön och berättelse om Bäfverns natur, hushållning och fångande." *Kungl. Svenska vetenskapsakademiens handlingar* 17 (1756): 207–221.

Greenberg, Joel. *A Feathered River across the Sky: The Passenger Pigeon's Flight to Extinction*. New York: Bloomsbury USA, 2014.

Gurnell, J., A. M. Gurnell, D. Demeritt, P. W. W. Lurz, M. D. F. Shirley, S. P. Rushton, C. G. Faulkes, S. Nobert, and E. J. Hare. *The Feasibility and Acceptability of Reintroducing the European Beaver to England*. Natural England Commissioned Report NECR002. Peterborough: Natural England and the People's Trust for Endangered Species, 2009.

Harting, James Edmund. *British Animals Extinct within Historic Times*. Boston: J. R. Osgood, 1880.

Herman, William C. "The Last Passenger Pigeon." *Auk* 65, no. 1 (1948): 77–80.

Hill, John. *A History of the Materia Medica*. London: Longman, Histch, and Hawes, 1751.

Hodge, Clifton Fremont. "A Last Word on the Passenger Pigeon." *Auk* 29, no. 2 (1912): 169–175.

Hodge, Clifton Fremont. *Nature Study and Life*. Boston: Ginn & Co., 1902.

Hodge, Clifton Fremont. "The Passenger Pigeon Investigation." *Auk* 28, no. 1 (1911): 49–53.

Hoel, Adolf. "Moskusokser til Svalbard." *Norsk Jæger- og Fisker Forenings Tidsskrift* 58 (1929): 326–328.

Hoel, Adolf. "Overføring av moskusokser til Svalbard." *Norge: Tidsskrift om vårt land* 6, no. 1 (1930): 15–18.

Hone, Elisabeth. *The Present Status of the Muskox in Arctic North America and Greenland*. Cambridge, MA: American Committee for International Wild Life Protection, 1934.

Hornaday, William T. *The Extermination of the American Bison*. Washington, DC: US Government Printing Office, 1889.

Hornaday, William T. *Our Vanishing Wild Life: Its Extermination and Preservation*. New York: Charles Scribner's Sons, 1913.

Jackson, H. T. "Attitude in Conservation." In *Silent Wings: A Memorial to the Passenger Pigeon*, edited by Walter E. Scott, 19–24. Madison: Wisconsin Society for Ornithology, 1947.

Kosack, Joe. "Passenger Pigeon: *Ectopistes migratorius*." Brochure, Pennsylvania Game Commission, March 2010. http://www.pgc.pa.gov/Wildlife/EndangeredandThreatened/Pages/PassengerPigeon.aspx.

Landsforeningen. "Moskusoksen bør vernes på Grønland." *Norge tidskrift om vårt land* 3, no. 23 (February 1927): 406–407.

Leopold, Aldo. "On a Monument to the Pigeon." In *Silent Wings: A Memorial to the Passenger Pigeon*, edited by Walter E. Scott, 3–5. Madison: Wisconsin Society for Ornithology, 1947.

Leopold, Aldo. *A Sand County Almanac: And Sketches Here and There*. Oxford: Oxford University Press, 1968. First published 1949.

Ling, Ingvar. "Bävern i Norrbotten." *Norrbottens Natur* 37 (1981): 6–10.

Løno, Odd *Transplantation of Muskox in Europe and North-America*, Meddelelser No. 84. Oslo: Norsk Polarinstitutt, 1960.

Lydekker, Richard. *The Royal Natural History*, vol. 4. London: Frederick Warne & Co., 1895.

Mather, Cotton. Untitled. Printed in Arlie W. Schorger, "Unpublished Manuscripts by Cotton Mather on the Passenger Pigeon." *Auk* 55, no. 3 (1938): 471–477.

Macdonald, David W., Françoise H. Tattersall, Stephen Rushton, Andy B. South, Shaila Rao, Peter Maitland, and Rob Strachan. "Reintroducing the Beaver (*Castor fiber*) to Scotland: A Protocol for Identifying and Assessing Suitable Release Sites." *Animal Conservation* 3 (2000): 125–133.

Mershon, W. B. *The Passenger Pigeon*. New York: Outing Publishing, 1907.

Modin, Erik. "Anteckningar om bäfvern, dess förekomst och fångst m.
m. i Västerbotten under förra hälften af 1800-talet." *Svenska Jägarförbundets Nya Tidskrift* 45 (1907): 269–275.

Modin, Erik. "Bör ej nägot göras för bäfverns återinförade i vårt land?"
Svenska Jägareförbundets Tidskrift 49 (1911): 192–194.

Nilsson, S. *Skandinavisk Fauna*. Vol. 1, *Däggdjuren*. Lund: Gleerups, 1847.

Nordiska Museet, *Skansens zoologiska trädgård: Kort vägledning för
besökande*. Stockholm: Centraltryckeriet, 1903.

Office of Communications, Department of the Interior. "A Passing in
Cincinnati, September 1, 1914." Washington, DC: US Government
Printing Office, 1976.

Olstad, O., and P. Tuff. "Innplanting av moskusokser på Dovrefjell." In
Årsmelding om Det Norkse Skogvesen, 51–55. Oslo: Skogdirektøren, 1942.

Pliny the Elder. *The Natural History*. 6 vols. Translated by John Bostock
and H. T. Riley. London: Henry G. Bohn, 1855–1857.

Post, G. H. von. "På besök hos Västerbottens bävrar." *Västerbottens läns
jaktvårdsförening årsbok* (1930): 49–66.

Ringstrand, Nils G. "Jaktvård." *Västerbottens läns jaktvårdsförening årsbok*
(1921): 5–7.

Russell, John. *Boke of Nurture*, edited by Frederick J. Furnivall. Bungay,
UK: John Childs and Son, 1867.

Salvesen, Sigvald. "The Beaver in Norway." *Journal of Mammalogy* 9
(1928): 99–104.

Scott, Walter E. "Dedication." In *Silent Wings: A Memorial to the Passenger Pigeon*, edited by Walter E. Scott, 1. Madison: Wisconsin Society for
Ornithology, 1947.

Scottish National Heritage. *Re-introduction of the European beaver to Scotland: A Public Consultation*. Perth: Scottish Natural Heritage, 1998.

Shufeldt, R. W. "Anatomical and Other Notes on the Passenger Pigeon (*Ectopistes migratorius*) Lately Living in the Cincinnati Zoölogical Gardens." *Auk* 32, no. 1 (1915): 29–41.

Smart, Gregory. *Birds on the British List: Their Title to Enrollment Considered*. London: R. H. Porter, 1886.

Spielberg, Steven, dir. *Jurassic Park*. Universal City, CA: Universal Pictures, 1993.

Stefansson, Vilhjalmur. *The Northward Course of Empire*. New York: Harcourt, Brace and Co., 1922.

Steffens, Haagen Krog. "Nicolai Benjamin Aall." In *Slægten Aall*, 430–432. Centraltrykkeriet: Kristiania, 1908.

Svenonius, Fredr. "Bjurälfdalens karstlandskap i norra Jämtland." *Sveriges Natur* 1 (1910): 73–80.

Swederus, G. *Skandinaviens Jagt: Djurfänge och Vildafugl*. Stockholm: P. A. Norstedt & Söner, 1832.

Sylvén, A. "Bävern tillbaka till Västerbotten." *Västerbottens läns jaktvårdsförening årsbok* (1922): 38–39.

Sylvén, A. "Våra bävrar." *Västerbottens läns jaktvårdsförening årsbok* (1924): 7–9.

Sylvén, A., Gunnar Beronius, and Nils Almlöf. "Till våra lärare!" *Västerbottens läns jaktvårdsförening årsbok* (1921): 1.

Sylvén, A., and Lennart Wahlberg. "Årsmötet 1924." *Västerbottens läns jaktvårdsförening årsbok* (1924): 69–70.

Taflin, Carin. "Mina år med Eric Festin." *Jämten* 79 (1986): 92–102.

Unander, F. "Ett från svenska jagtbanan försvunnet dyrbart djur." *Svenska Jägarförbundets Nya Tidskrift* 11 (1873): 28–33.

Vogel, Heinz. *Das abenteuerliche Leben des JJ1 alias Bruno alias Beppo. Ein Bilderbuch für Erwachsene und Kinder*. Hohenems: Bucher, 2013.

Wahlberg, Lennart. "Bäverns återbördande till Syd-Lappland." *Västerbottens läns jaktvårdsförening årsbok* (1925): 6–23.

Wenninger, F. J. "The Passenger Pigeon." *American Midland Naturalist* 1, no. 8 (1910): 227–228.

Winter, Johan. "Et och annat om Bäfvern." *Svenska Jägarförbundets Nya Tidskrift* 11 (1873): 110–112.

Online Primary Sources

Direktoratet for Naturforvalting, www.dirnat.no.

EU LIFE Arctos project, http://www.life-arctos.it/.

Long Now Foundation, Revive & Restore, http://longnow.org/revive/.

Rewilding Europe, https://www.rewildingeurope.com.

TEDxDeExtinction event, March 13, 2013, Washington DC, http://long now.org/revive/events/tedxdeextinction.

Todd McGrain, artist, Lost Bird Project, http://www.toddmcgrain.com.

EU LIFE Ursus project, http://www.pnab.it/natura-e-territorio/orso/life -ursus.html.

Secondary Literature

Alagona, Peter S. *After the Grizzly: Endangered Species and the Politics of Place in California.* Berkeley: University of California Press, 2013.

Albrecht, Glenn, Gina-Maree Sartore, Linda Connor, Nick Higgin-botham, Sonia Freeman, Brian Kelly, Helen Stain, Anne Tonna, and Georgia Pollard. "Solastalgia: The Distress Caused by Environmental Change." *Australasian Psychiatry* 15, no. S1 (2007): S95–S98.

Anderson, Warwick. "Climates of Opinion: Acclimatization in Nineteenth-Century France and England." *Victorian Studies* 35, no. 2 (1992): 135–157.

Assmann, Jan. "Communicative and Cultural Memory." In *Cultural Memory Studies: An International and Interdisciplinary Handbook*, edited by A. Eril and A. Nünning, 109–118. Berlin: Walter de Gruyter, 2008.

Assmann, Jan. *Moses the Egyptian: The Memory of Egypt in Western Monotheism*. Cambridge, MA: Harvard University Press, 1997.

Averill, James R., George Catlin, and Kynm Koo Chon. *Rules of Hope*. New York: Springer Science and Business Media, 1990.

Balaguer, Luis, Adrián Escudero, José F. Martín-Duque, Ignacio Mola, and James Aronson. "The Historical Reference in Restoration Ecology: Re-defining a Cornerstone Concept." *Biological Conservation* 176 (2014): 12–20.

Barrow, Mark V. Jr. *Nature's Ghosts: Confronting Extinction from the Age of Jefferson to the Age of Ecology*. Chicago: University of Chicago Press, 2009.

Barrow, Mark V. Jr. "Teetering on the Brink of Extinction: The Passenger Pigeon, the Bison, and American Zoo Culture in the Late Nineteenth and Early Twentieth Centuries." In *The Ark and Beyond: The Evolution of Zoo and Aquarium Conservation*, edited by Ben A. Minteer, Jane Maienschein, and James P. Collins, 51–64. Chicago: University of Chicago Press, 2018.

Benson, Etienne. "The Urbanization of the Eastern Gray Squirrel in the United States." *The Journal of American History* 100, no. 3 (2013): 691–710.

Biess, Frank. "Forum: History of Emotions." *German History* 28, no. 1 (2010): 67–80.

Bladow, Kyle, and Jennifer Ladino, eds. *Affective Ecocriticism: Emotion, Embodiment, Environment*. Lincoln: University of Nebraska Press, 2018.

Bloeser, Claudia, and Titus Stahl. "Hope." In *The Stanford Encyclopedia of Philosophy* (Spring 2017 edition), edited by Edward N. Zalta. Metaphysics Research Lab, Stanford University. https://plato.stanford.edu/archives/spr2017/entries/hope/.

Bourke, Joanna. *Fear: A Cultural History*. London: Virago, 2005.

Bristow, Tom. "Memory." *Environmental Humanities* 5 (2014): 307–311.

Broomhall, Susan, ed. *Early Modern Emotions: An Introduction*. London: Routledge, 2017.

Bryant, William. "Whole System, Whole Earth: The Convergence of Ecology and Technology in Twentieth Century American Culture." PhD diss., University of Iowa, May 2006.

Burnett, Melissa S., and Dale A. Lunsford. "Conceptualizing Guilt in the Consumer Decision-Making Process." *Journal of Consumer Marketing* 11, no. 3 (1994): 33–43.

Cafaro, Philip, and Richard Primack. "Species Extinction Is a Great Moral Wrong." *Biological Conservation* 170 (2014): 1–2.

Charmaz, Kathy, and Melinda J. Milligan. "Grief." In *Handbook of the Sociology of Emotions*, edited by J. E. Stets and J. H. Turner, 516–543. Boston: Springer, 2006.

Chase, Malcolm, and Christopher Shaw. "The Dimensions of Nostalgia." In *The Imagined Past: History and Nostalgia*, edited by Christopher Shaw and Malcolm Chase, 1–17. Manchester: Manchester University Press, 1989.

Chew, Matthew, and Andrew Hamilton. "The Rise and Fall of Biotic Nativeness: A Historical Perspective." In *Fifty Years of Invasion Ecology: The Legacy of Charles Elton*, edited by David M. Richardson, 35–47. Oxford: Blackwell Publishing, 2011.

Chrulew, Matthew. "Reversing Extinction: Restoration and Resurrection in the Pleistocene Rewilding Projects." *Humanimalia* 2, no. 2 (2011): 4–27.

Clavero, Miguel. "Shifting Baselines and the Conservation of Non-native Species." *Conservation Biology* 28, no. 5 (2014): 1434–1436.

Clewell, Andre F., and James Aronson. *Ecological Restoration: Principles, Values, and Structure of an Emerging Profession*. 2nd ed. Washington, DC: Island Press, 2013.

Coates, Peter. "Creatures Enshrined: Wild Animals as Bearers of Heritage." *Past and Present* 226, no. S10 (2015): 272–298.

Coleman, Jon T. *Vicious: Wolves and Men in America*. New Haven, CT: Yale University Press, 2004.

Coles, Bryony. *Beavers in Britain's Past*. Oxford: Oxbow Books, 2006.

Court, Franklin E. *Pioneers of Ecological Restoration: The People and Legacy of the University of Wisconsin Arboretum*. Madison: University of Wisconsin Press, 2012.

Cronon, William. *Nature's Metropolis: Chicago and the Great West*. New York: W. W. Norton, 1991.

Cronon, William. "The Trouble with Wilderness, or Getting Back to the Wrong Nature." In *Uncommon Ground: Rethinking the Human Place in Nature*, edited by William Cronon, 69–90. New York: W. W. Norton & Co., 1995.

Crutzen, Paul J. "Geology of Mankind." *Nature* 415 (January 3, 2012): 23.

Crutzen, Paul J., and Eugene F. Stoermer. "The 'Anthropocene.'" *Global Change Newsletter* 41 (2000): 17–18.

Cunsolo, Ashlee, and Neville R. Ellis. "Ecological Grief as a Mental Health Response to Climate Change-Related Loss." *Nature Climate Change* 8 (2018): 275–281.

Cunsolo, Ashlee, and Karen Landman, eds. *Mourning Nature: Hope at the Heart of Ecological Loss and Grief*. Montreal: McGill-Queen's University Press, 2017.

Decker, Stephen E., Alistair J. Bath, Alvin Simms, Uwe Lindner, and Edgar Reisinger. "The Return of the King or Bringing Snails to the Garden? The Human Dimensions of a Proposed Restoration of European Bison (*Bison bonasus*) in Germany." *Restoration Ecology* 18, no. 1 (2010): 41–51.

Dehler, Gregory J. *The Most Defiant Devil: William Temple Hornaday and His Controversial Crusade to Save American Wildlife*. Charlottesville: University of Virginia Press, 2013.

Deinet, S., C. Ieronymidou, L. McRae, I. J. Burfield, R. P. Foppen, B. Collen, and M. Böhm. *Wildlife Comeback in Europe: The Recovery of Selected Mammal and Bird Species*. Final report to Rewilding Europe by ZSL, BirdLife International, and the European Bird Census Council. London: ZSL, 2013.

Donlan, C. Josh, Joel Berger, Carl E. Bock, Jane H. Bock, David A. Burney, James A. Estes, Dave Foreman, Paul S. Martin, Gary W. Roemer, Felisa A. Smith, Michael E. Soulé, and Harry W. Greene. "Pleistocene Rewilding: An Optimistic Agenda for Twenty-First Century Conservation." *American Naturalist* 168, no. 5 (November 2006): 660–681.

Donlan, C. Josh, and Harry W. Greene. "NLIMBY: No Lions in My Backyard." In *Restoration and History: The Search for a Usable Environmental Past*, edited by Marcus Hall, 293–308. New York: Routledge, 2010.

Donlan, C. Josh, Harry W. Greene, Joel Berger, Carl E. Bock, Jane H. Bock, David A. Burney, James A. Estes, Dave Foreman, Paul S. Martin, Gary W. Roemer, Felisa A. Smith, and Michael E. Soulé. "Re-wilding North America." *Nature* 436 (2005): 913–914.

Driessen, Clemens, and Jamie Lorimer. "Back-Breeding the Aurochs: The Heck Brothers, National Socialism and Imagined Geographies for Nonhuman Lebensraum." In *Hitler's Geographies*, edited by P. Giaccaria and C. Minca, 138–157. Chicago: University of Chicago Press, 2016.

Dunlap, Thomas R. "Remaking the Land: The Acclimatization Movement and Anglo Ideas of Nature." *Journal of World History* 8, no. 2 (1997): 303–319.

Dunlap, Thomas R. *Saving America's Wildlife*. Princeton, NJ: Princeton University Press, 1988.

Dunlap, Thomas R. "Sport Hunting and Conservation, 1880–1920." *Environmental History Review* 12 (1988): 51–60.

Egan, Michael. "Chemical Unknowns: Preliminary Outline for an Environmental History of Fear." In *Framing the Environmental Humanities*, edited by Hannes Bergthaller and Peter Mortensen, 124–138. Leiden: Brill, 2018.

Ferguson, Mark A., and Nyla R. Branscombe. "Collective Guilt Mediates the Effect of Beliefs about Global Warming on Willingness to Engage in Mitigation Behavior." *Journal of Environmental Psychology* 30 (2010): 135–142.

Fiege, Mark. *The Republic of Nature: An Environmental History of the United States.* Seattle: University of Washington Press, 2012.

Fletcher, Amy L. "Bring 'em Back Alive: Taming the Tasmanian Tiger Cloning Project." *Technology in Society* 30 (2008): 194–201.

Fletcher, Amy L. "Genuine Fakes: Cloning Extinct Species as Science and Spectacle." *Politics and the Life Science* 20, no. 1 (2010): 48–60.

Fletcher, Amy L. *Mendel's Ark: Biotechnology and the Future of Extinction.* Dordrecht: Springer Science, 2014.

Flores, Dan. *American Serengeti: The Last Big Animals of the Great Plains.* Lawrence: University Press of Kansas, 2016.

Folch, J., M. J. Cocero, P. Chesné, J. L. Alabart, V. Domínguez, Y. Cognié, A. Roche, A. Fernández-Árias, J. I. Martí, P. Sánchez, E. Echegoyen, J. F. Beckers, A. Sánchez Bonastre, and X. Vignon. "First Birth of an Animal from an Extinct Subspecies (*Capra pyrenaica pyrenaica*) by Cloning." *Theriogenology* 71 (2009): 1026–1034.

Fountain, Steven M. "Ranchers' Friend and Farmers' Foe: Reshaping Nature with Beaver Reintroduction in California." *Environmental History* 19 (2014): 239–269.

Fowler, Hayden. "Epilogue: New World Order—Nature in the Anthropocene." In *Animals in the Anthropocene: Critical Perspectives on Non-Human Futures,* edited by Human Animal Research Network Editorial Collective, 243–254. Sydney: Sydney University Press, 2015.

Frawley, Jodi, and Iain McCalman, eds. *Rethinking Invasion Ecologies from the Environmental Humanities.* Abingdon: Routledge, 2014.

Freedman, Jonathan L., Sue A. Wallington, and Evelyn Bless. "Compliance without Pressure: The Effect of Guilt." *Journal of Personality and Social Psychology* 7 (1967): 117–124.

Friese, Carrie. "Cloning in the Zoo: When Zoos Become Parents." In *The Ark and Beyond: The Evolution of Zoo and Aquarium Conservation*, edited by Ben A. Minteer, Jane Maienschein, and James P. Collins, 267–278. Chicago: University of Chicago Press, 2018.

Freud, Sigmund. "Trauer under Melancholie" [Mourning and melancholia]. *Internationale Zeitschrift für Ärztliche Psychoanalyse* 4 (1917): 288–301.

Gaynor, Andrea. "Environmental History and the History of Emotions." Histories of Emotion research group. https://historiesofemotion.com/2017/06/16/environmental-history-and-the-history-of-emotions/.

Gibson, Katherine, Deborah Bird Rose, and Ruth Fincher, eds. *Manifesto for Living in the Anthropocene*. New York: Punctum Books, 2015.

Godfrey, Joseph J. *A Philosophy of Human Hope*. Dordrecht: Marinus Nijhoff, 1987.

Goedeke, T. L., and S. Rikoon. "Otters as Actors: Scientific Controversy, Dynamism of Networks, and the Implications of Power in Ecological Restoration." *Social Studies of Science* 38, no. 1 (2008): 111–132.

Greene, Harry W. "Pleistocene Rewilding and the Future of Biodiversity." In *After Preservation: Saving American Nature in the Age of Humans*, edited by Ben A. Minteer and Stephen J. Pyne, 105–113. Chicago: University of Chicago Press, 2015.

Halbwachs, Maurice. *On Collective Memory*. Chicago: University of Chicago Press, 1992. First published 1925.

Hall, Marcus. *Earth Repair: A Transatlantic History of Environmental Restoration*. Charlottesville: University of Virginia Press, 2005.

Hall, Marcus. "Restoration and the Search for Counter-narratives." In *The Oxford Handbook for Environmental History*, edited by Andrew C. Isenberg, 309–331. Oxford: Oxford University Press, 2014.

Halley, Duncan, and Frank Rosell. "Population and Distribution of European Beavers (*Castor fiber*)." *Lutra* 46, no. 3 (2003): 91–101.

Handeland, Kjell, Torstein Tengs, Branko Kokotovic, Turid Vikøren, Roger D. Ayling, Bjarne Bergsjø, Ólöf G. Sigurðardóttir, and Tord Bretten. "*Mycoplasma ovipneumoniae*: A Primary Cause of Severe Pneumonia Epizootics in the Norwegian Muskox (*Ovibos moschatus*) Population." *PLoS One* 9, no. 9 (2014): e106116.

Hardenberg, Wilko Graf von. "Another Way to Preserve: Hunting Bans, Biosecurity and the Brown Bear in Italy, 1930–60." In *The Nature State: Rethinking the History of Conservation*, edited by Wilko Graf von Hardenberg, Matthew Kelly, Claudia Leal, and Emily Wakild, 55–75. Abingdon: Routledge, 2017.

Hartman, Göran. "The Beaver (*Castor fiber*) in Sweden." In *Restoring the European Beaver: 50 Years of Experience*, edited by Göran Sjöberg and John P. Ball, 13–17. Sofia-Moscow: Pensoft, 2011.

Head, Lesley. "The Anthroposceneans." *Geographical Research* 53, no. 3 (2015): 313–320.

Head, Lesley. *Hope and Grief in the Anthropocene: Re-conceptualising Human-Nature Relations*. Abingdon: Routledge, 2016.

Heise, Ursula K. *Imagining Extinction: The Cultural Meanings of Endangered Species*. Chicago: University of Chicago Press, 2016.

Heise, Ursula K. "Lost Dogs, Last Birds, and Listed Species: Cultures of Extinction." *Configurations* 18 (2010): 49–72.

Higgs, Eric. *Nature by Design: People, Natural Process, and Ecological Restoration*. Cambridge, MA: MIT Press, 2003.

Hobbs, Richard J. "Grieving for the Past and Hoping for the Future: Balancing Polarizing Perspectives in Conservation and Restoration." *Restoration Ecology* 21, no. 2 (2013): 145–148.

Hobbs, Richard J., Eric S. Higgs, and Carol Hall. *Novel Ecosystems: Intervening in the New Ecological World Order*. Oxford: Wiley-Blackwell, 2013.

Horn, Susanne, Walter Durka, Ronny Wolf, Aslak Ermala, Annegret Stubbe, Michael Stubbe, and Michael Horfeiter. "Mitochondrial Genomes Reveal Slow Rates of Molecular Evolution and the Timing of

Speciation in Beavers (Castor), One of the Largest Rodent Species." *PLOS One* 6, no. 1 (2011): e14622.

Hume, Julian P. "Large-Scale Live Capture of Passenger Pigeons *Ectopistes migratorius* for Sporting Purposes: Overlooked Illustrated Documentation." *Bulletin of the British Ornithologists' Club* 135 (2015): 174–184.

Isenberg, Andrew C. *The Destruction of the Bison: An Environmental History, 1750–1920.* Cambridge: Cambridge University Press, 2000.

Isenberg, Andrew C. "The Returns of the Bison: Nostalgia, Profit, and Preservation." *Environmental History* 2, no. 2 (1997): 179–196.

Ivakhiv, Adrian J. *Ecologies of the Moving Image: Cinema, Affect, Nature.* Waterloo, Canada: Wildred Laurier University Press, 2013.

Jacoby, Karl. *Crimes against Nature: Squatters, Poachers, Thieves, and the Hidden History of American Conservation.* Berkeley: University of California Press, 2001.

Jones, Ryan Tucker. *Empire of Extinction: Russians and the North Pacific's Beasts of the Sea, 1741–1867.* Oxford: Oxford University Press, 2014.

Jordan, William R. III, and George Lubick. *Making Nature Whole: A History of Ecological Restoration.* Washington, DC: Island Press, 2011.

Jørgensen, Dolly. "After None: Memorialising Animal Species Extinction through Monuments." In *Animals Count: How Population Size Matters in Animal-Human Relations*, edited by Nancy Cushing and Jodi Frawley, 183–199. Abingdon: Routledge, 2018.

Jørgensen, Dolly. "Beastly Belonging in the Premodern North." In *Visions of North in Premodern Europe*, edited by Dolly Jørgensen and Virginia Langum, 183–205. Turnhout: Brepols, 2018.

Jørgensen, Dolly. "Migrant Muskox and the Naturalization of National Identity in Scandinavia." In *The Historical Animal*, edited by Susan Nance, 184–201. Syracuse, NY: Syracuse University Press, 2015.

Jørgensen, Dolly. "Muskox in a Box and Other Tales of Containers as Domesticating Mediators in Animal Relocation." In *Animal Housing and Human-Animals Relations: Politics, Practices and Infrastructures*, edited by

Tone Druglitrø and Kristian Bjørkdahl, 100–114. Abingdon: Routledge, 2016.

Jørgensen, Dolly. "Not by Human Hands: Five Technological Tenets for Environmental History in the Anthropocene." *Environment and History* 20 (2014): 479–489.

Jørgensen, Dolly. "Presence of Absence, Absence of Presence and Extinction Narratives." In *Nature, Temporality and Environmental Management*, edited by Lesley Head, Katarina Saltzman, Gunhild Setten, and Marie Stensek, 45–58. Abingdon: Routledge, 2017.

Jørgensen, Dolly. "Rethinking Rewilding." *Geoforum* 65 (2015): 482–488.

Jørgensen, Dolly. "What's History Got to Do with It? A Response to Seddon's Definition of Reintroduction." *Restoration Ecology* 19, no. 6 (2011): 705–708.

Jørgensen, Dolly. "Who's the Devil? Species Extinction and Environmentalist Thought in Star Trek." In *Star Trek and History*, edited by Nancy Reagin, 242–259. New York: Wiley & Sons, 2013.

Karlsen, Jan. "Moskus, Zibet, Castoreum og Ambra: Animalske droger i det gamle apotek." *Cygnus—en norsk farmahistorisk skriftserie* 10 (October 2004): 39–47.

Katz, Eric. "Another Look at Restoration: Technology and Artificial Nature." In *Restoring Nature: Perspectives from the Social Sciences and Humanities*, edited by Paul H. Gobster and R. Bruce Hull, 37–48. Washington, DC: Island Press, 2000.

Knoll, Martin. "Hunting in the Eighteenth Century: An Environmental History Perspective." *Historical Social Research* 29 (2004): 9–36.

Kolbert, Elizabeth. "How to Write about a Vanishing World." *New Yorker*, October 15, 2018. https://www.newyorker.com/magazine/2018/10/15/how-to-write-about-a-vanishing-world.

Konigsberg, Ruth David. *The Truth about Grief: The Myth of Its Five Stages and the New Science of Loss.* New York: Simon and Schuster, 2011.

Konstan, David. *Pity Transformed.* London: Duckworth, 2001.

Kretz, Lisa. "Hope in Environmental Philosophy." *Journal of Agricultural and Environmental Ethics* 26, no. 5 (2013): 925–944.

Kurin, Richard. *The Smithsonian's History of America in 101 Objects*. New York: Penguin Press, 2013.

Kübler-Ross, Elisabeth. *On Death and Dying*. New York: Macmillan, 1969.

Kübler-Ross, Elisabeth, and David Kessler. *On Grief and Grieving: Finding the Meaning of Grief through the Five Stages of Loss*. New York: Scribner, 2005.

Ladino, Jennifer. *Reclaiming Nostalgia: Longing for Nature in American Literature*. Charlottesville: University of Virginia Press, 2012.

Lange, John H. "Bruno the Bear." *Indoor Built Environment* 15, no. 5 (2006): 391–392.

Langston, Nancy. "Restoration in the American National Forests: Ecological Processes and Cultural Landscapes." In *The Conservation of Cultural Landscapes*, edited by Mauro Agnoletti, 163–173. Wallingford, UK: CAB International, 2006.

Lent, Peter. *Muskoxen and Their Hunters: A History*. Norman: University of Oklahoma Press, 1999.

Light, Andrew. "Ecological Restoration and the Culture of Nature: A Pragmatic Perspective." In *Restoring Nature: Perspectives from the Social Sciences and Humanities*, edited by Paul H. Gobster and R. Bruce Hull, 49–70. Washington, DC: Island Press, 2000.

Loo, Tina. *States of Nature: Conserving Canada's Wildlife in the Twentieth Century*. Vancouver: UBC Press, 2006.

Lorimer, Jamie. *Wildlife in the Anthropocene: Conservation after Nature*. Minneapolis: Minnesota University Press, 2015.

Lorimer, Jamie, and Clemens Driessen. "Bovine Biopolitics and the Promise of Monsters in the Rewilding of Heck Cattle." *Geoforum* 48 (2013): 249–259.

Lorimer, Jamie, and Clemens Driessen. "From 'Nazi Cows' to Cosmopolitan 'Ecological Engineers': Specifying Rewilding through a History of Heck Cattle." *Annals of the American Association of Geographers* 106, no. 3 (2016): 631–52.

Lorimer, Jamie, and Clemens Driessen. "Wild Experiments at the Oostvaardersplassen: Rethinking Environmentalism in the Anthropocene." *Transactions of the Institute of Royal Geographers* 39, no. 2 (2014): 169–181.

Lowenthal, David. "Natural and Cultural Heritage." *International Journal of Heritage Studies* 11, no. 1 (2005): 81–92.

Lowenthal, David. "Nostalgia Tells It like It Wasn't." In *The Imagined Past: History and Nostalgia*, edited by Christopher Shaw and Malcolm Chase, 18–32. Manchester: Manchester University Press, 1989.

Lowenthal, David. *The Past Is a Foreign Country*. Cambridge: Cambridge University Press, 1985.

Macfarlane, Daniel. "Emotional and Environmental History at Niagara Falls." Otter blog, Network in Canadian History and Environment. http://niche-canada.org/2017/09/28/emotional-and-environmental-history-at-niagara-falls/.

Marsh, George Perkins. *Man and Nature, or Physical Geography as Modified by Human Action*. New York: Charles Scribner, 1864.

Merchant, Carolyn. *Reinventing Eden: The Fate of Nature in Western Culture*. New York: Routledge, 2003.

Minteer, Ben A. "When Extinction Is a Virtue." In *After Preservation: Saving American Nature in the Age of Humans*, edited by Ben A. Minteer and Stephen J. Pyne, 96–104. Chicago: University of Chicago Press, 2015.

Minton, Ann P., and Randall L. Rose. "The Effects of Environmental Concern on Environmentally Friendly Consumer Behavior: An Exploratory Study." *Journal of Business Research* 40, no. 1 (1997): 37–48.

Mkono, Muchazondida. "'Eating the Animals You Come to See': Tourists' Meat-Eating Discourses in Online Communicative Texts." In *Animals and Tourism: Understanding Diverse Relationships*, edited by Kevin Markwell, 211–225. Bristol: Channel View Publications, 2015.

Monbiot, George. *Feral: Searching for Enchantment on the Frontiers of Rewilding*. London: Penguin, 2013.

Mooallem, Jon. *Wild Ones: A Sometimes Dismaying, Weirdly Reassuring Story about Looking at People Looking at Animals in America*. New York: Penguin, 2013.

Mumford, Lewis. *Technics and Civilization*. Chicago: University of Chicago Press, 2010. First published 1934.

Noonan, Thomas S., and Roman K. Kovalev. "'The Furry 40s': Packaging Pelts in Medieval Northern Europe." In *States, Societies, Cultures. East and West: Essays in Honor of Jaroslaw Pelenski*, edited by Janusz Duzinkiewicz, 653–682. New York: Ross Publishing, 2004.

O'Gorman, Emily. "Belonging." *Environmental Humanities* 5 (2014): 283–286.

Osborne, Michael A. "Acclimatizing the World: A History of the Paradigmatic Colonial Science." *Osiris*, 2nd series, 15 (2001): 135–151.

Paddle, Robert. *The Last Tasmanian Tiger: The History and Extinction of the Thylacine*. Cambridge: Cambridge University Press, 2000.

Papworth, S. K., J. Rist, L. Coad, and E. J. Milner-Gulland. "Evidence for Shifting Baseline Syndrome in Conservation." *Conservation Letters* 2 (2009): 93–100.

Pauly, Daniel. "Anecdotes and the Shifting Baseline Syndrome of Fisheries." *Trends in Ecology & Evolution* 10, no. 10 (1995): 430.

Pedersen, Alwin. *Der Moschusochs*. Wittenberg Luterstadt: A. Ziemsen, 1958.

Pike, Sarah. "Mourning Nature: The Work of Grief in Radical Environmentalism." *Journal for the Study of Religion, Nature, and Culture* 10, no. 4 (2016): 419–441.

Poliquin, Rachel. *Beaver*. London: Reaktion Books, 2015.

Poliquin, Rachel. *The Breathless Zoo: Taxidermy and the Cultures of Longing*. University Park: Pennsylvania State University Press, 2012.

Powell, Miles. *Vanishing America: Species Extinction, Racial Peril, and the Origins of Conservation.* Cambridge, MA: Harvard University Press, 2016.

Price, Jennifer. *Flight Maps: Adventures with Nature in Modern America.* New York: Basic Books, 1999.

Pucek, Zdzisław, Irena P. Belousova, Zbigniew A. Krasiński, Małgorzata Krasińska, and Wanda Olech. *European Bison (Bison bonasus): Current State of the Species and Strategy for Its Conservation.* Strasbourg: Council of Europe Publishing, 2004.

Ritvo, Harriet. "Going Forth and Multiplying: Animal Acclimatization and Invasion." *Environmental History* 17, no. 2 (2012): 404–414.

Roberts, Peder, and Dolly Jørgensen. "Animals as Instruments of Norwegian Imperial Authority in the Interwar Arctic." *Journal for the History of Environment and Society* 1 (2016): 65–87.

Ronda, Margaret. "Mourning and Melancholia in the Anthropocene." *Post 45*, October 6, 2013. http://post45.research.yale.edu/2013/06/mourning-and-melancholia-in-the-anthropocene/.

Rose, Deborah Bird, Thom van Dooren, and Matthew Chrulew, eds. *Extinction Studies: Stories of Time, Death, and Generations.* New York: Columbia University Press, 2017.

Rosell, Frank, Orsolya Bozsér, Peter Collen, and Howard Parker. "Ecological Impact of Beavers *Castor fiber* and *Castor canadensis* and Their Ability to Modify Ecosystems." *Mammal Review* 35, no. 3–4 (2005): 248–276.

Rosen, Tatjana, and Alistair Bath. "Transboundary Management of Large Carnivores in Europe: From Incident to Opportunity." *Conservation Letters* 2 (2009) 109–114.

Rosenwein, Barbara H., ed. *Anger's Past: The Social Uses of an Emotion in the Middle Ages.* Ithaca, NY: Cornell University Press, 1989.

Rosenwein, Barbara H. *Emotional Communities in the Early Middle Ages.* Ithaca, NY: Cornell University Press, 2006.

Rosenwein, Barbara H. "Problems and Methods in the History of Emotions." *Passions in Context* 1 (2010): 1–32.

Rotherham, Ian D., and Robert A. Lambert, eds. *Invasive and Introduced Plants and Animals: Human Perceptions Attitudes and Approaches to Management*. London: Earthscan, 2011.

Sandler, Ronald. "The Ethics of Reviving Long Extinct Species." *Conservation Biology* 28, no. 2 (2013): 354–360.

Sandler, Ronald. "Techno-Conservation in the Anthropocene: What Does It Mean to Save a Species?" In *The Routledge Companion to the Environmental Humanities*, edited by Ursula Heise, Jon Christensen, and Michelle Niemann, 72–81. Abingdon: Routledge, 2017.

Schama, Simon. *Landscape and Memory*. Toronto: Random House, 1995.

Scheer, Monique. "Are Emotions a Kind of Practice (And Is That What Makes Them Have a History)? A Bourdieuian Approach to Understanding Emotion." *History and Theory* 51 (2012): 193–220.

Schepers, Frans, and Paul Jepson. "Rewilding in a European Context." *International Journal of Wilderness* 22, no. 2 (2016): 25–30.

Seddon, Philip J. "From Reintroduction to Assisted Colonization: Moving Along the Conservation Translocation Spectrum." *Reintroduction Ecology* 18, no. 6 (2010): 796–802.

Seegert, Natasha. "Queer Beasts: Ursine Punctures in Domesticity." *Environmental Communication: A Journal of Nature and Culture* 8, no. 1 (2014): 75–91. https://doi.org/10.1080/17524032.2013.798345.

Seymour, Nicole. "Toward an Irreverent Ecocriticism." *Journal of Ecocriticism* 4, no. 2 (2012): 56–71.

Shade, Patrick. *Habits of Hope: A Pragmatic Theory*. Nashville: Vanderbilt University Press, 2001.

Shade, Patrick. "Shame, Hope and the Courage to Transgress." In *Theories of Hope: Exploring Alternative Affective Dimensions of Human Experience*, edited by Rochelle Green, 47–70. London: Rowman & Littlefield, 2019.

Shapiro, Beth. *How to Clone a Mammoth: The Science of De-extinction*. Princeton, NJ: Princeton University Press, 2015.

Shell, Hanna Rose. "Skin Deep: Taxidermy, Embodiment and Extinction in W. T. Hornaday's Buffalo Group." *Proceedings of the California Academy of Sciences* 55, S1, no. 5: 88–112.

Siipi, Helena, and Leonard Finkelman. "The Extinction and De-extinction of Species." *Philosophy and Technology* 30, no. 4 (2017): 427–441.

Small, Neil. "Theories of Grief: A Critical Review." In *Grief, Mourning and Death Ritual*, edited by Jennifer L. Hockey, Jeanne Katz, and Neil Small, 19–48. London: Open University Press, 2001.

Smith, Bruce D., and Melinda A. Zeder. "The Onset of the Anthropocene." *Anthropocene* 4 (2013): 8–13.

Smith, Laura. "On the 'Emotionality' of Environmental Restoration: Narratives of Guilt, Restitution, Redemption and Hope." *Ethics, Policy & Environment* 17 (2014): 286–307.

Society for Ecological Restoration International Science & Policy Working Group. *The SER International Primer on Ecological Restoration*. Tucson: Society for Ecological Restoration International, 2004.

Songster, E. Elena. *Panda Nation: The Construction and Conservation of China's Modern Icon*. Oxford: Oxford University Press, 2018.

Stafford, William. "'This Once Happy Country': Nostalgia for Premodern Society." In *The Imagined Past: History and Nostalgia*, edited by Christopher Shaw and Malcolm Chase, 33–46. Manchester: Manchester University Press, 1989.

Steffen, Will, Paul J. Crutzen, and John R. McNeill. "The Anthropocene: Are Humans Now Overwhelming the Great Forces of Nature?" *Ambio* 36, no. 8 (2007): 614–621.

Stearns, Peter N., and Carol Z. Stearns. "Emotionology: Clarifying the History of Emotions and Emotional Standards." *American Historical Review* 90, no. 4 (1985): 813–836.

Steinberg, Ted. *Down to Earth: Nature's Role in American History*. Oxford: Oxford University Press, 2002.

Stine, Jeffrey K. "Placing Environmental History on Display." *Environmental History* 7, no. 4 (2002): 566–588.

Street, Paul. "Painting Deepest England: The Late Landscapes of John Linnell and the Uses of Nostalgia." In *The Imagined Past: History and Nostalgia*, edited by Christopher Shaw and Malcolm Chase, 68–80. Manchester: Manchester University Press, 1989.

Stroebe, Margaret, Henk Schut, and Kathrin Boerner. "Cautioning Health-Care Professionals: Bereaved Persons Are Misguided through the Stages of Grief." *OMEGA—Journal of Death and Dying* 74, no. 4 (2017): 455–473.

Swart, Sandra. "Zombie Zoology: History and Reanimating Extinct Animals." In *The Historical Animal*, edited by Susan Nance, 54–71. Syracuse, NY: Syracuse University Press, 2015.

Tamm, Marek. "Beyond History and Memory: New Perspectives in Memory Studies." *History Compass* 11, no. 6 (2013): 458–473.

Throop, William. "The Rationale for Environmental Restoration." In *The Ecological Community: Environmental Challenges for Philosophy, Politics, and Morality*, edited by Roger S. Gottleib, 39–55. New York: Routledge, 1997.

Trigger, David, Jane Mulcock, Andrea Gaynor, and Yann Toussaint. "Ecological Restoration, Cultural Preferences and the Negotiation of 'Nativeness' in Australia." *Geoforum* 39 (2008): 1273–1283.

Tuan, Yi-Fu. *Landscapes of Fear*. New York: Pantheon Book, 1979.

Turner, Stephanie S. "Open-Ended Stories: Extinction Narratives in Genome Time." *Literature and Medicine* 26, no. 1 (2007): 55–82.

van Dooren, Thom. *Flight Ways: Life and Loss at the Edge of Extinction*. New York: Columbia University Press, 2014.

van Dooren, Thom. "Invasive Species in Penguin Worlds: An Ethical Taxonomy of Killing for Conservation." *Conservation and Society* 9, no. 4 (2011): 286–298.

Vera, Frans. "The Shifting Baseline Syndrome in Restoration Ecology." In *Restoration and History: The Search for a Usable Environmental Past*, edited by Marcus Hall, 98–110. New York: Routledge, 2009.

Walker, Brett L. *The Lost Wolves of Japan*. Seattle: University of Washington Press, 2005.

Walton, Todd, and Wendy S. Shaw. "Living with the Anthropocene Blues." *Geoforum* 60 (2015): 1–3.

Waterworth, Jayne M. *A Philosophical Analysis of Hope*. New York: Palgrave Macmillan, 2004.

Weik von Mossner, Alexa. *Affective Ecologies: Empathy, Emotion, and Environmental Narrative*. Columbus: Ohio State University Press, 2017.

Weik von Mossner, Alexa, ed. *Moving Environments: Affect, Emotion, Ecology, and Film*. Waterloo, Canada: Wilfrid Laurier University Press, 2014.

Welzer, Harald. "Communicative Memory." In *Cultural Memory Studies: An International and Interdisciplinary Handbook*, edited by A. Eril and A. Nünning, 285–300. Berlin: Walter de Gruyter, 2008.

White, Richard. "From Wilderness to Hybrid Landscapes: The Cultural Turn in Environmental History." *Historian* 66, no. 3 (2004): 557–564.

Wigh, Bengt. "Animal Bones from the Viking Town of Birka, Sweden." In *Leather and Fur: Aspects of Early Medieval Trade and Technology*, edited by Esther Cameron, 81–90. London: Archetype Publications, 1998.

Wilson, E. O. *Biophilia*. Cambridge, MA: Harvard University Press, 1984.

Woodward, C. Vann. "Review of *The Past Is a Foreign Country*." *History and Theory* 26, no. 3 (1987): 346–352.

Zedler, Joy B. "Ecological Restoration: The Continuing Challenge of Restoration." In *The Essential Aldo Leopold: Quotations and Commentaries*, edited by Curt Meine and Richard L. Knight, 116–126. Madison: University of Wisconsin Press, 1999.

Zimmer, Carl. "Bringing Them Back to Life." *National Geographic* (April 2013). http://ngm.nationalgeographic.com/2013/04/125-species-revival/zimmer-text.

Index

Affect, 12

American bison (*Bison bison*), 16, 102, 104, 127, 150

Åmli (Norway), 38, 40, 139

Andersson, Bengt, 78–80

Angaard, Jon, 56, 74

Anger, 99–103

Anthropocene, 5, 150–151

Arbman, Sven, 23, 39–46

Art, 131–132. *See also* ATM (street artist); Malling, Sverre; McGrain, Todd; Ruthven, John A.

ATM (street artist), 1–2, 24, 147

Audubon, John James, 53, 93–95, 97, 115

Bear, European brown (*Ursus arctos*)
books about, 127–128
in Germany, 119–121, 124–125
in Italy, 124
remembrance of, 125–128

Beaver, European (*Castor fiber*). *See also* Guilt

in Britain, 1, 139–140
decline of, 30–31
description of, 25–26
in Germany, 33
history of human relations with, 26–28
hunting of, 27
in medieval sources, 27–29
memories of, 140–143
reintroduction to Sweden of, 23–54, 141
reproduction of, 16
safari ,141
stories about, 34–36, 140–141

Beaver, North American (*Castor canadensis*), 26

Behm, Alarik, 29–30, 32–34, 36, 39

Belonging
of beavers in Sweden, 32, 36, 52–53
as cultural category, 8–9
as motivator for restoration, 4
of muskoxen in Norway, 69, 85–86